Cambridge Lower Secondary

Science

STAGE 9: WORKBOOK

Aidan Gill, Heidi Foxford,
Dorothy Warren

Collins

William Collins' dream of knowledge for all began with the publication of his first book in 1819.

A self-educated mill worker, he not only enriched millions of lives, but also founded a flourishing publishing house. Today, staying true to this spirit, Collins books are packed with inspiration, innovation and practical expertise. They place you at the centre of a world of possibility and give you exactly what you need to explore it.

Collins. Freedom to teach.

Published by Collins
An imprint of HarperCollins*Publishers*
The News Building
1 London Bridge Street
London
SE1 9GF

Browse the complete Collins catalogue at
www.collins.co.uk

© HarperCollins*Publishers* Limited 2018

10 9 8 7 6 5 4

ISBN 978-0-00-825473-5

MIX
Paper from
responsible sources
FSC www.fsc.org **FSC™ C007454**

This book is produced from independently certified FSC paper to ensure responsible forest management.

For more information visit:
www.harpercollins.co.uk/green

British Library Cataloguing in Publication Data
A catalogue record for this publication is available from the British Library.

Authors: Aidan Gill, Heidi Foxford, Dorothy Warren
Development editors: Elizabeth Barker, Peter Batty, Richard Needham
Team leaders: Mark Levesley, Peter Robinson, Aidan Gill
Commissioning project manager: Carol Usher
Commissioning editors: Joanna Ramsay, Rachael Harrison
In-house editor: Natasha Paul
Copyeditor: Tony Clappison
Proofreader: Sara Hulse
Answer checker: Mike Smith
Illustrator: Jouve India Private Limited
Cover designer: Gordon MacGilp
Cover illustrations: Maria Herbert-Liew
Internal designer: Jouve India Private Limited
Typesetter: Jouve India Private Limited
Production controller: Tina Paul
Printed and bound by: CPI Group (UK) Ltd, Croydon CR0 4YY
All test-style questions and sample answers in these resources were written by the authors. In Cambridge tests, the way marks are awarded may be different.

Acknowledgements

The publishers gratefully acknowledge the permission granted to reproduce the copyright material in this book. Every effort has been made to trace copyright holders and to obtain their permission for the use of copyright material. The publishers will gladly receive any information enabling them to rectify any error or omission at the first opportunity. p174: Science History Images/ Alamy Stock Photo

Contents

How to use this book

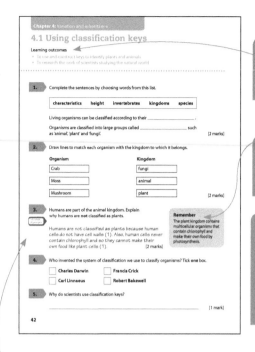

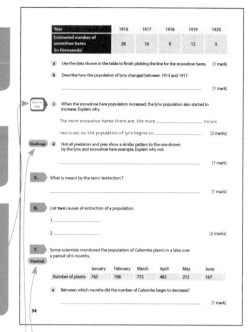

The outcomes show what you will cover in the questions

There are handy hints and tips in the 'remember' boxes

Learn how to structure your answers with 'show me' questions where part of the answer is already completed for you

The worked examples will show you a sample answer and how the marks are awarded

Stretch yourself with these challenge questions

There are questions developing your practical skills throughout

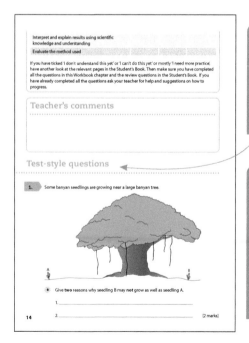

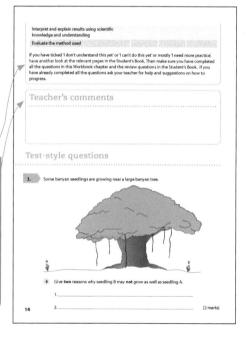

At the end of the chapter, try the test-style questions! The questions will cover the topics within the chapter

At the end of the chapter, fill in the table to work out the areas you understand and the areas where you might need more practice. There is space for your teacher to comment too

Biology

1.1 The photosynthesis equation

Learning outcomes

- To model photosynthesis, using a word equation
- To learn about different types of variables
- To learn about repeatability and reliability
- To learn about doing trial runs

1. Complete the sentences by choosing words from this list.

| chloroplasts | heat | light | protein | sugar |

Plants use energy carried by _____ from the Sun.

They use this energy to make _____. [2 marks]

2. What do plants need to grow well? Tick **two** boxes.

☐ air   ☐ insects   ☐ flowers   ☐ water   [2 marks]

3. Why do plants **not** need to eat other plants or animals? Tick the best reason.

☐ they do not need any nutrients   ☐ they do not have a digestive system

☐ they can make their own food   ☐ they get everything they need from the soil

[1 mark]

4. The diagram shows a plant cell from a leaf.

a Which letter shows a chloroplast?

_____ [1 mark]

b What is the name of the green pigment

in chloroplasts? _____

[1 mark]

c Why is this pigment important? Tick **two** boxes.

☐ it is food for the plant   ☐ it absorbs light energy from the Sun

☐ it helps the plant to make food   ☐ it stops the plant losing water   [2 marks]

5. Explain why a leaf contains cells packed with chloroplasts.

Worked Example

Leaves are the site of photosynthesis. (1)

Many cells in a leaf are packed with chloroplasts because these are the site of photosynthesis. (1)   [2 marks]

Remember
Chlorophyll and chloroplasts are two key words that look similar, but are different. Chloroplasts CONTAIN the green pigment chlorophyll.

6. A potato is a **tuber**. This is the part of the potato plant that grows underground. A tuber stores nutrients. Khalid wants to find out if potato tuber cells contain chloroplasts.

a Name a piece of apparatus Khalid should use to find out if potato tuber cells contain chloroplasts.

_____   [1 mark]

Show Me

b Khalid discovers that potato tuber cells do not contain chloroplasts.

Explain why potato tuber cells do **not** contain chloroplasts.

These are needed for photosynthesis, which requires

_____ *(1); there is no light* _____ . *(1)* [2 marks]

7. Giant aroid plants grow on the floor of tropical rainforests in Borneo. Their leaves can measure 3 metres across.

Explain why giant aroid plants have such large leaves.

_____

_____

_____   [2 marks]

8. Complete the word equation for photosynthesis by choosing words from this list.

| oxygen | air | carbon dioxide | water | carbon | minerals |

_____ + _____ → glucose + _____

[3 marks]

9. Explain why photosynthesis is described as an **endothermic** reaction.

_____ [1 mark]

10. Stomata are tiny holes found on leaves. Name **one** substance needed for photosynthesis that enters a plant through the stomata.

_____ [1 mark]

11. **a** List **two** ways in which a plant uses the glucose it makes.

1._____

2._____ [2 marks]

b How is unused glucose stored in a plant?

_____ [1 mark]

12. Eimear says: 'Without plants, energy cannot be passed along a food chain.'
Do you agree with this statement? Explain your answer.

Challenge

_____

_____ [2 marks]

13. Annika is investigating how the distance away from a light source affects the rate of photosynthesis in pondweed. She changes the intensity of light by varying the distance of the pondweed from the lamp. Pondweed gives off bubbles of oxygen when it is photosynthesising.

Practical

Annika sets up the apparatus shown below.

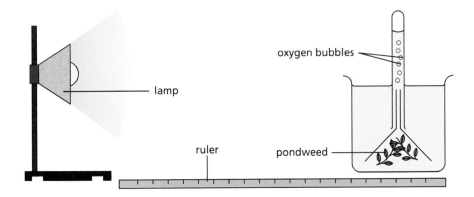

a Complete the table to show how Annika should carry out her investigation.
Tick **one** box in each **row**.

Factor	Variable to be changed (independent)	Variable to be kept the same (control)	Variable to be measured (dependent)
Type of pondweed			
Distance of beaker from lamp			
Time taken to count number of bubbles			
Number of bubbles counted			

[4 marks]

b Annika's results contain an anomalous result. What is meant by the term 'anomalous result'?

_____ [1 mark]

c Suggest **one** cause of the anomalous result.

_____ [1 mark]

1.2 Water and mineral salts

Learning outcomes
- To explain the importance of water and mineral salts for plant growth
- To describe how water, mineral salts and sugars are transported in a plant
- To describe the importance of some elements found in mineral salts

1. The diagram shows the parts of a tomato plant.

   a Name the part of the plant that absorbs water.

   _____ [1 mark]

   b Name the part containing hollow tubes that transport water to all parts of the plant.

   _____ [1 mark]

   c What are these hollow tubes called?

   _____ [1 mark]

   d The stem contains tubes that transport sugars. What is the name of these tubes?

   _____ [1 mark]

leaf

stem

root

2. Draw lines to match each word to its meaning.

Word	Meaning
Transpiration	A cell that is limp because there is little water inside the cell
Flaccid	A cell that is swollen because it is full of water
Turgid	Movement of water up through a plant

[2 marks]

3. Describe how water vapour escapes from a plant.

Show Me

The water vapour escapes from the _____ through tiny holes called

_____ . The water vapour moves from the inside of the leaf to the

_____ by a process called _____ . [2 marks]

4. Complete the sentences by choosing words from this list.

deficiency	efficiency	leaves	phloem	roots

Plants need mineral salts in order to grow healthily. Mineral salts enter the plant

through its _____ .

If plants do not get the elements that they need, they have _____

symptoms.   [2 marks]

5. Chen has a potted plant. He waters it every day but some leaves are turning yellow.

Chen looks at this table to find out what the problem might be.

Name of the element from a mineral salt	Why the plant needs the element
nitrogen (from nitrate salts)	to make proteins so that the plant can grow
magnesium (from magnesium salts)	to make chlorophyll
phosphorus (from phosphate salts)	for respiration and growth

a Chen's plant does not have enough of **one** of these elements from mineral salts. Use the information in the table to suggest which element it is.

_____   [1 mark]

b What could Chen do to stop the leaves turning yellow? Tick **one** box.

☐ add fertiliser to the soil   ☐ spray the leaves with water

☐ add worms to the soil   ☐ put the plant in a smaller pot   [1 mark]

6.

Challenge

Gabriella's pumpkin plant is wilting. She has not watered it for a week. Explain why the plant wilts without water.

_____

_____

_____

> **Remember**
> Plant cells store water in their vacuoles. If a plant cell has a lot of water, high pressure inside that cell pushes outwards. The cells push against each other giving the plant support so it stays upright.

[2 marks]

Nasreen is investigating the effect of mineral salts on plant growth.

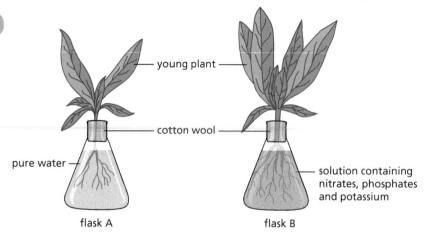

She puts the flasks in a sunny position for four weeks. The diagrams above show the plants after four weeks.

a Explain why Nasreen grew one plant in pure water.

_____

_____ [1 mark]

b Suggest **one** way in which Nasreen could measure the growth of the plants.

_____

_____ [1 mark]

Nasreen's results are shown below.

Flask	Observations
A	The plant did not grow well The edges of leaves turned brown
B	The plant grew well The leaves stayed green

c What could she do to make her results more reliable?

_____

_____ [1 mark]

d What conclusion can Nasreen make from her results?

_____

_____ [1 mark]

e Nasreen read that if plants lack phosphates, they develop purple spots on their leaves.

Suggest **one** reason to explain why she did **not** observe any purple spots.

_____ [1 mark]

Self-assessment

Tick the column which best describes what you know and what you are able to do.

What you should know:	I don't understand this yet	I need more practice	I understand this
Photosynthesis can be modelled using a word equation: carbon dioxide + water → glucose + oxygen			
Photosynthesis requires energy from light. Inside chloroplasts, chlorophyll traps some energy from light			
Gases (such as carbon dioxide, oxygen and water vapour) diffuse in and out of open stomata			
Plants use water for photosynthesis and for support			
Plants need many elements contained in minerals – including nitrogen, phosphorus, potassium and magnesium – for healthy growth			

You should be able to:	I can't do this yet	I need more practice	I can do this by myself
Select ideas and produce plans for testing based on previous knowledge			
Make careful observations and sufficient measurements to reduce error and make results reliable			
Draw conclusions			
Decide whether to use information collected first hand or from secondary sources			

Interpret and explain results using scientific
knowledge and understanding

Evaluate the method used

If you have ticked 'I don't understand this yet' or 'I can't do this yet' or mostly 'I need more practice', have another look at the relevant pages in the Student's Book. Then make sure you have completed all the questions in this Workbook chapter and the review questions in the Student's Book. If you have already completed all the questions ask your teacher for help and suggestions on how to progress.

Teacher's comments

Test-style questions

1. Some banyan seedlings are growing near a large banyan tree.

A   B

a Give **two** reasons why seedling B may **not** grow as well as seedling A.

1. _____

2. _____ [2 marks]

b Why is the banyan tree called a producer? Tick **one** box.

☐ because it produces many leaves

☐ because it produces its own food

☐ because it takes in food to produce wood

☐ because it reproduces quickly [1 mark]

c Give **two** reasons why banyan trees and other plants need water.

1. _____

2. _____ [2 marks]

d The banyan tree needs mineral salts for healthy growth.

Name **one** mineral salt that plants need for healthy growth.

_____ [1 mark]

e Describe how mineral salts enter the plant.

_____ [1 mark]

f Describe how glucose is made in the leaves of the banyan tree.

_____

_____

_____ [3 marks]

2.

Safia puts a cut celery stem in a beaker of red-coloured water.

The leaves turn red when the water reaches them.

Safia investigates if the air temperature affects how quickly the water gets to the leaves.

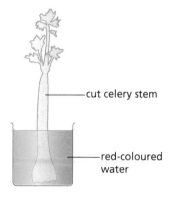

cut celery stem

She measures how long it takes for the leaves to turn red at two different air temperatures. Each time she uses a new piece of celery.

red-coloured water

She does this three times for each temperature.

a   What is the **independent** variable in this investigation?

_____   [1 mark]

b   What is the **dependent** variable?

_____   [1 mark]

c   Name **one** variable that Safia should control.

_____   [1 mark]

d   Explain why Safia repeated her measurements three times at each temperature.

_____   [1 mark]

e   Explain what happens to water after it moves up the stem of a plant.

_____

_____

_____   [3 marks]

2.1 Flowers and pollination

Learning outcomes

- To recall the parts of a flower and their functions
- To compare and contrast the different ways in which pollination occurs
- To explain how pollination leads to fertilisation, and to the production of seeds and fruits

1. What is the main function of a flower? Tick the best answer.

☐ to make food for the plant ☐ to absorb water and nutrients

☐ to contain the sexual organs of a plant ☐ to provide nectar for insects [1 mark]

2. Complete the labels in the diagram using these words:

stigma	ovary	style	anther

filament

ovule

[3 marks]

3. Name **two** parts of a flower that are male.

_____ _____ [2 marks]

4. What do plants produce that contains the male gametes?

_____ [1 mark]

5. What is the function of the:

a stigma _____ [1 mark]

b ovary? _____ [1 mark]

6. Complete the sentences by choosing words from this list.

| nectar | seeds | eggs | pollen | roots | oxygen |

Pollination is an important part of the life cycle of plants.

_____ is transferred between flowering plants,

allowing the plants to reproduce and make _____ . [2 marks]

7. Describe the difference between **self-pollination** and **cross-pollination**.

Worked Example

Self-pollination is when the pollen is transferred to a stigma in the same flower, or a flower on the same plant (1).

Cross-pollination is when pollen is transferred to a different plant of the same species (1). [2 marks]

> **Remember**
> Pollination is **not** the same as fertilisation. Pollination is the **transfer of pollen** from an anther to a stigma. Fertilisation is the **joining of the nuclei** from the gametes inside a pollen grain and an ovule.

8. Which of the following is a disadvantage of cross-pollination?

- [ ] It gives lots of variety within a plant species.
- [ ] It makes the plant more at risk of disease.
- [ ] It needs wind or animals to transfer the pollen.
- [ ] It does not need wind or animals to transfer the pollen. [1 mark]

9. Pollen can be transferred from one plant to another by insects. List **two** features of insect-pollinated flowers.

1. _____

2. _____ [2 marks]

10. Wind can also transfer pollen from plant to plant. Explain how the following characteristics help wind-pollinated plants.

Show Me

a Large amount of pollen: *Because most of the pollen does not* _____

b Anthers hang outside the flower: *so* _____ *directly into the wind.* [2 marks]

11. The sentences below describe what happens in fertilisation in a plant. The sentences are in the wrong order. Write the numbers 1-4 next to the sentences to show the **correct order** in which they happen. The first has been done for you.

The nucleus inside a pollen grain and the nucleus in an ovule join up

A pollen grain lands on the stigma **1**

A pollen tube grows down to the ovary

The ovary develops into a fruit and each ovule becomes a seed containing an embryo   [1 mark]

12. A pomegranate plant produces fruits, each containing about 600 seeds.
Suggest why a large number of seeds is an advantage for the plant.

Challenge

_____   [1 mark]

13. Chetan and Devi are investigating how sugar affects the growth of pollen tubes.

Practical   **Chetan's plan:**

A   Put some pollen into a concentrated sugar solution.

B   Put some pollen into a dilute sugar solution.

C   After one hour, estimate which solution has more pollen tubes.

Devi's plan:

A   Get four beakers:

- three contain different concentrations of sugar solution: 5%, 10% and 15%

- one contains pure water.

B   Add the same amount of pollen to each of the different sugar solutions.

C   After one hour, count how many pollen tubes can be seen.

a   Explain **two** reasons why Devi's plan is better than Chetan's.

1._____

_____

2._____

_____   [2 marks]

b   Name a piece of apparatus that you might use to observe the growth of the pollen tubes.

_____   [1 mark]

c The table shows Devi's results:

Sugar concentration (%)	5	10	15
Number of pollen cells with tube growth	25	35	45

Devi concludes that 'increasing the sugar concentration increases the number of pollen tubes that can be seen'.

Do Devi's results support his conclusion? Explain your answer using evidence from the table.

_____

_____ [2 marks]

2.2 Seed dispersal

Learning outcomes

- To recall some of the different ways in which flowering plants disperse their seeds
- To explain why plants disperse their seeds
- To describe what happens during germination
- To evaluate a method

1. Complete the sentences by choosing words from this list.

seed	fruit	flower	root	gamete

After fertilisation each ovule becomes a _____ .

In many plants, the ovary develops into a _____ . [2 marks]

2. A seed contains an embryo. What does the embryo grow into? Tick **one** box.

☐ The shoot and root of a new plant   ☐ A food store

☐ A tough protective outer layer of the seed   ☐ Another seed   [1 mark]

3. Explain why plants disperse their seeds.

Show Me

Seed dispersal reduces competition for things such as

_____ from _____.

[2 marks]

Remember
The word 'disperse' means 'to spread out'. So seed dispersal happens when plants spread out their seeds.

4. The diagrams show some different seeds.

Copy and complete the table by naming the type of seed dispersal. Then describe the features that allow each seed to be dispersed in this way. The first one has been done for you.

Type of seed	Type of seed dispersal	Features of seed that help dispersal
dandelion	wind	Seed has parachute that catches the wind
sandbur		
tomato		

dandelion

sandbur

tomato

[4 marks]

5. The sea bean grows on the banks of rivers in Africa, Australia and South America. Its seeds form inside large woody pods. The pods drop into the river and float away.

a Name the type of seed dispersal used by the sea bean.

_____ [1 mark]

b Explain why sea bean seeds may be found on beaches many miles from the parent plant.

_____

_____ [2 marks]

6. Explain what is meant by 'germination'.

_____ [1 mark]

7. List **three** resources needed for successful germination.

1._____

2._____

3._____ [3 marks]

8.

Challenge

The diagram shows the life cycle of a tomato plant.

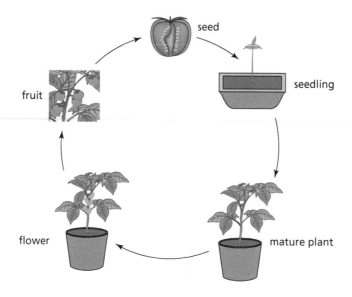

Describe the processes that take place in the life cycle of a tomato plant. Use the diagram to help you.

_____

_____ [2 marks]

9.

Practical

Moninder wants to find out the best temperature for mustard seeds to germinate.

He uses three Petri dishes. Each dish has a layer of moist tissue paper covered with 30 mustard seeds.

Moninder puts one dish in a warm cupboard, one in a cool cupboard and one in a refrigerator. He waters all of the seeds regularly for 21 days. His results for the first four days are shown in the table.

	Number of germinated mustard seeds			
	Day 1	Day 2	Day 3	Day 4
Warm cupboard	0	5	20	24
Cool cupboard	0	2	8	14
Refrigerator	0	0	0	0

a How many seeds had germinated by day 4 in the warm cupboard?

_____ [1 mark]

b What conclusion can you draw from Moninder's results? Tick **one** box.

☐ Mustard seeds need mineral salts to germinate

☐ Fewer seeds germinate in warm conditions

☐ More mustard seeds germinate in cooler conditions

☐ More mustard seeds germinate in warm conditions [1 mark]

c Explain why **no** seeds germinated in the refrigerator.

_____ [1 mark]

d Liam does a similar investigation in a laboratory. He uses apparatus that keeps the mustard seeds at set temperatures. He observes how many seeds germinate at 10, 15, 20 and 25 °C.

Explain why Liam's method is better at finding out the 'best temperature' for mustard seeds to germinate.

_____

_____ [2 marks]

Self-assessment

Tick the column which best describes what you know and what you are able to do.

What you should know:	I don't understand this yet	I need more practice	I understand this
Pollination is the transfer of pollen from the anther to the stigma in a flower of the same species			
Fertilisation happens when a pollen grain nucleus joins with a female gamete nucleus in an ovule			
The fertilised ovule develops into a seed and the rest of the carpel forms a fruit surrounding the seed			
Plant species disperse their seeds in many different ways including by wind, water, animals and explosion			
You should be able to:	I can't do this yet	I need more practice	I can do this by myself
Make observations			

Interpret and explain results using scientific knowledge and understanding			
Present conclusions and evaluation of working methods			
Compare the results and methods used by others			
Decide which measurements and observations are necessary and what apparatus to use			
Discuss the importance of evidence			

If you have ticked 'I don't understand this yet' or 'I can't do this yet' or mostly 'I need more practice', have another look at the relevant pages in the Student's Book. Then make sure you have completed all the questions in this Workbook chapter and the review questions in the Student's Book. If you have already completed all the questions ask your teacher for help and suggestions on how to progress.

Teacher's comments

....

Test-style questions

....

1. The diagram (right) shows the flower of a lily plant.

    a Write the names of the structures labelled **A** and **B**.

    A: _____

    B: _____ [2 marks]

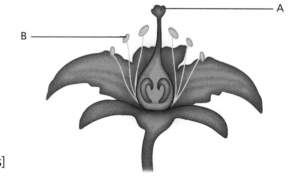

b What is the function of the part labelled **B**?

_____ [1 mark]

c This lily is pollinated by insects. Which of the following are common characteristics of insect-pollinated plants? Tick **two** boxes.

☐ No scent   ☐ Large brightly coloured petals

☐ Produce nectar   ☐ Feathery stigma   [2 marks]

d Explain how insects are able to pollinate a plant.

_____

_____

_____ [2 marks]

e The seeds of this lily are light, flat and slightly winged. Use this information to suggest how the seeds are dispersed.

Method of dispersal: _____

Reason you think this: _____

_____

_____ [2 marks]

f Explain why it is an advantage to the lily's offspring to disperse its seeds.

_____

_____

_____ [2 marks]

g The life cycle of the lily involves pollination, fertilisation and germination.

Draw a line to match each word with its correct meaning.

Word	Meaning
Pollination	When the embryo inside a seed starts to grow
Fertilisation	The joining of the male and female gamete nuclei
Germination	Transfer of pollen grains from the stamen to the stigma

[2 marks]

h Explain why dry lily seeds kept in a seed packet do **not** germinate.

_____ [1 mark]

2. Imran has three different seeds, each with a different shape. He wants to investigate which seed will be dispersed over the furthest distances.

He drops each seed from a height of 2 metres and measures how far the seed travels horizontally.

Seed	Distance travelled (cm)		
	Trial 1	Trial 2	Trial 3
A	74	48	68
B	3	6	4
C	0	1	1

a Use the results to predict which seed is normally dispersed by the wind.
Explain your answer.

_____ [1 mark]

b Why did Imran do **three** trials for each seed?

_____ [1 mark]

c How could Imran make his results more accurate? Tick **one** box only.

☐ Measure the distance travelled in millimetres

☐ Drop each seed from 3 metres

☐ Investigate more seeds

☐ Measure the distance travelled by each seed once

[1 mark]

3.1 Food webs and energy flow

Learning outcomes

- To identify the different types of feeding relationship in a habitat
- To model energy flow through different trophic levels in a habitat, using food chains and food webs
- To explain the importance of decomposers in a habitat

1. The figures below show four living things found in a grassland habitat.

snake   grass   hawk   grasshopper

Not to scale

- grasshoppers eat grass
- snakes eat grasshoppers
- hawks eat snakes

Complete this food chain for the four living things shown.

Grass ⟶ _____ ⟶ _____ ⟶ _____   [1 mark]

2. Look at the organisms in question 1. Complete the table below by giving the names of **one** predator, **one** prey, **one** producer and **one** primary consumer.

Predator	Prey	Producer	Primary consumer

[4 marks]

3. Which type of organism uses the Sun's energy to make its own food?

☐ **consumer**   ☐ **producer**   ☐ **herbivore**   ☐ **predator**   [1 mark]

4. Draw lines to match each word with its meaning.

Word	Meaning
Herbivore	animal that eats both plants and animals
Carnivore	animal that eats plants
Omnivore	animal that eats other animals

[2 marks]

5. A food chain shows 'what eats what'.

What do the arrows in a food chain represent?

_____ [1 mark]

6. Juan says: 'All the energy in a food chain comes from the Sun.'

Worked Example

Explain why this statement is true.

Energy from the Sun is stored in the biomass made
by plants during photosynthesis (1). This energy
is passed along the food chain when one organism
consumes another (1). [2 marks]

> **Remember**
> 'Biomass' is the mass
> of living material that
> organisms make for
> themselves. The biomass
> of an organism contains a
> store of energy.

7. The diagram below shows a food web.

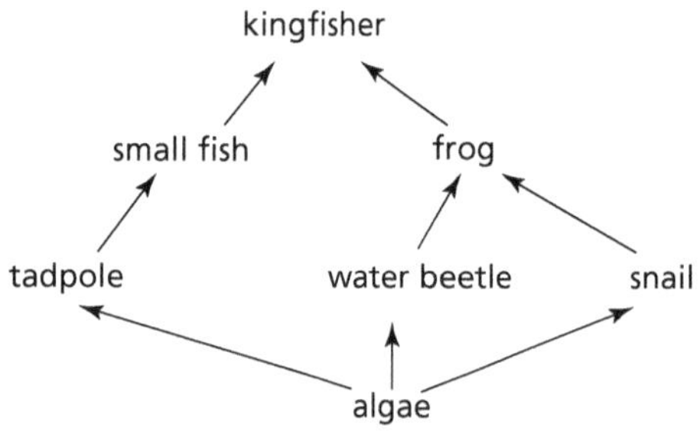

a Use the food web to draw a food chain that has **four** trophic levels.

_____ → _____ → _____ → _____ [1 mark]

Show Me

b All the frogs in the food web die of a disease. Explain what will happen
to the size of the population of snails.

The number of snails will _____ because

_____ . [2 marks]

8. **Not all** of the biomass from one organism will get to the animal at the next
trophic level. Explain why.

Challenge

_____

_____ [2 marks]

28

9. Tick the boxes to indicate **two** decomposers.

☐ viruses ☐ bacteria

☐ fungi ☐ spiders [2 marks]

10. Explain how decomposers help plants.

_____

_____ [2 marks]

11.

Practical

Petra is investigating the growth of mould on pieces of bread left in different conditions. She puts a slice of bread in each of two plastic bags.

• One bag is left in a refrigerator.

• One bag is left in a warm room.

Petra counts the number of mould colonies (spots) on each piece of bread after 7 days and 14 days.

a Draw a table for Petra to record her observations. [1 mark]

b What sort of graph or chart would be best for Petra to present her results?

_____ [1 mark]

3.2 Adaptation and survival

Learning outcomes

- To recall the resources that plants and animals need
- To explain ways in which land organisms are adapted to their habitats
- To explain ways in which water organisms are adapted to their habitats

1. Complete the table to show what resources organisms need to survive.
   Add **one** tick to each row. The first row has been done for you.

Resource	Plants	Animals	Both plants and animals
Sunlight	✓		
Water			
A source of food			
Carbon dioxide			

[3 marks]

2. Complete the sentence by choosing words from this list.

temperature   habitat   reproduce   move   survive   respire

An adaptation is a special characteristic that allows an organism to _____

and _____ in a particular _____ .   [3 marks]

3. Draw lines to match each physical adaptation of a polar bear to its
   correct explanation of how the feature helps.

Adaptation   **How the feature helps**

| Thick layer of fat | | to catch prey |

| Sharp claws | | provides camouflage from prey |

| White fur | | provides insulation from the cold | [2 marks]

4. The barrel cactus grows in the desert.

   This plant is adapted to life in the desert. Explain how
   each adaptation helps the barrel cactus to survive.

30

a Large swollen stem:

stores water so the cactus _____. [1 mark]

b Thick spines rather than leaves:

_____

_____. [2 marks]

5. Many animals hibernate during the winter. Explain how this helps them to survive.

_____

_____ [2 marks]

6.

Challenge

Scientists observe a group of fennec foxes in the Sahara Desert. The foxes hunt at night and rest in burrows during the day.

Suggest how these behaviours could help the fennec fox survive in the desert.

Remember

When a question asks you to 'suggest', think about what you already know and come up with your own ideas about the information you are given.

_____

_____ [2 marks]

7.

Practical

Rajiv wants to find out which conditions maggots prefer to live in. He uses a large container that has four different areas. Each area has a different condition to the others, as shown in the figure below:

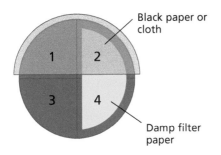

Area 1: dark and dry
Area 2: dark and moist
Area 3: light and dry
Area 4: light and moist

The maggots can move easily between the different areas.

a Write a plan that describes how Rajiv could find out which condition maggots prefer.

_____

_____

_____

_____ [4 marks]

b What type of graph or chart would be most suitable for displaying these results?

_____ [1 mark]

3.3 Populations and extinction

Learning outcomes

- To describe some factors that affect population size
- To describe some human influences on habitats

1. Complete the sentences by choosing words from this list.

temperature	competition	predation	water availability

The size of a population can be changed by living factors or non-living (physical) factors.

Non-living factors include _____ , sunlight, _____

and pollution. Living factors include disease, _____ and

_____ . [2 marks]

2. Name **one** physical factor that could limit a plant population at the bottom of a lake.

_____ [1 mark]

3. The graph shows how the human population has changed since 1800.

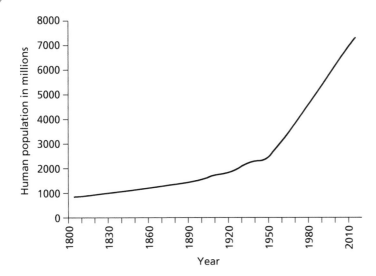

Which **two** of the following are possible reasons for the changes in the human population size? Tick **two** boxes.

☐ **better access to clean water**   ☐ **better healthcare**

☐ **more disease**   ☐ **less food available**   [2 marks]

4. The graph shows how the estimated population of snowshoe hares and lynx in one part of Canada changed over 20 years.

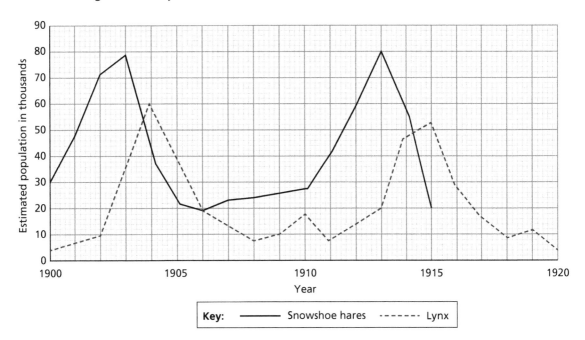

Year	1916	1917	1918	1919	1920
Estimated number of snowshoe hares (in thousands)	28	16	9	12	5

a Use the data shown in the table to finish plotting the line for the snowshoe hares. [1 mark]

b Describe how the population of lynx changed between 1913 and 1917.

_____ [1 mark]

Show Me

c When the snowshoe hare population increased, the lynx population also started to increase. Explain why.

The more snowshoe hares there are, the more _____ the lynx

have to eat, so the population of lynx begins to _____. [2 marks]

Challenge **d** Not all predators and prey show a similar pattern to the one shown by the lynx and snowshoe hare example. Explain why not.

_____ [1 mark]

5. What is meant by the term 'extinction'?

_____ [1 mark]

6. List **two** causes of extinction of a population.

1. _____

2. _____ [2 marks]

7. Some scientists monitored the population of *Cabomba* plants in a lake over a period of 6 months.

Practical

	January	February	March	April	May	June
Number of plants	765	768	772	483	212	167

a Between which months did the number of *Cabomba* begin to decrease?

_____ [1 mark]

b List **two** possible reasons for this decrease.

1. _____

2. _____ [2 marks]

Challenge **c** The number of *Cabomba* plants in the lake was estimated by counting the number in one 'sample' area. Suggest **one** reason why the data may **not** be accurate.

_____

_____ [2 marks]

3.4 Human influences on the environment

Learning outcomes

- To describe ways in which humans damage habitats, and the reasons for it
- To describe ways in which humans alter the Earth and its atmosphere
- To explain some ways in which we can live more sustainably

1. An increase in the human population causes an increase in the demand for electricity.

Name **two** fossil fuels that are burned to generate electricity.

1. _____

2. _____ [2 marks]

2. Which of the following is a problem linked to an increase in the burning of fossil fuels? Tick **one** box.

☐ **global warming** ☐ **deforestation**

☐ **increased recycling** ☐ **increased marine biodiversity** [1 mark]

3. Describe the difference between renewable and non-renewable energy sources.

Show Me

A renewable energy source can be _____ .

A non-renewable energy source _____ .

[2 marks]

4. The rainforests in many countries are being destroyed by deforestation.

Give **one** reason why.

_____ [1 mark]

5. Explain how deforestation can affect the number of species found in an area.

_____

_____ [2 marks]

6. What is meant by the term 'sustainable development'?

_____ [1 mark]

7. Anastasia studies water pollution. She uses indicator species to find out how polluted the water is at **two** different sites in a town.

Species	Sensitivity to pollution
Stonefly larva	Cannot live in polluted water
Alderfly larva	Cannot live in polluted water
Freshwater mussel	Can live in slightly polluted water
Damselfly larva	Can live in slightly polluted water
Bloodworm	Can live in polluted water
Rat-tailed maggot	Can live in polluted water

Sample site	Species found
A	Alderfly larva, stonefly larva, freshwater mussel
B	Rat-tailed maggot, bloodworm

a Which sample came from a polluted pond?

_____ [1 mark]

b Anastasia takes a sample from another site and she only finds damselfly larvae. She says 'This shows that the site is polluted.'

Do you agree? Explain your answer.

_____

_____ [2 marks]

Self-assessment

Tick the column which best describes what you know and what you are able to do.

What you should know:	I don't understand this yet	I need more practice	I understand this
Food chains show how energy passes from one organism to the next in feeding relationships			
Food webs link together food chains in a habitat			
Organisms that are in the same position in a food chain are at the same trophic level			
Trophic levels are producers, primary consumers, secondary consumers, tertiary consumers			
Decomposers decay dead organisms and faeces, releasing mineral salts back into the soil			
Animals and plants have physical adaptations that help them to survive in specific habitats			
Many animals have behavioural adaptations that help them to survive, such as hibernating, migrating and being nocturnal			
Many factors cause species to become extinct including new diseases, competition for resources and catastrophic events			

	I can't do this yet	I need more practice	I can do this by myself
As the human population increases there is a greater demand for space and resources, including energy from fuels			
Humans use deforestation, intensive farming and quarrying to create more space, grow more food and gather more resources			
Air pollution causes problems such as global warming and acid rain			
Indicator species show pollution levels in the air and in water			

You should be able to:	I can't do this yet	I need more practice	I can do this by myself
Explain and interpret results using scientific knowledge and understanding			
Evaluate the methods used			
Draw conclusions			
Look critically at sources of secondary data			
Describe patterns seen in results			
Decide which measurements and observations are necessary			
Select ideas and produce plans for testing based on previous knowledge			

If you have ticked 'I don't understand this yet' or 'I can't do this yet' or mostly 'I need more practice', have another look at the relevant pages in the Student's Book. Then make sure you have completed all the questions in this Workbook chapter and the review questions in the Student's Book. If you have already completed all the questions ask your teacher for help and suggestions on how to progress.

Teacher's comments

Test-style questions

1. The diagram shows a food chain for organisms that live in an area of farmland.

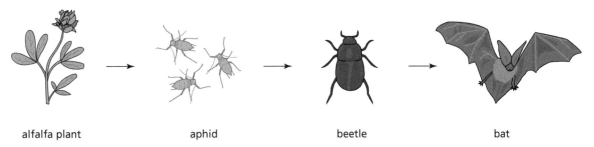

| alfalfa plant | aphid | beetle | bat |

a Complete the table to show whether each of the organisms is a predator, prey or both. Add **one** tick to each row. The first one has been done for you.　　[2 marks]

Animal	Predator	Prey	Both
aphid		✓	
beetle			
bat			

b Name a herbivore shown in the food chain.

_____ [1 mark]

c In this food chain, the alfalfa plant is the producer. What is meant by 'producer'?

_____ [1 mark]

d In summer, a large number of birds migrate to the farmland and eat large numbers of beetles.

Suggest why this might increase the size of the aphid population.

_____ [1 mark]

e Many microorganisms live in the soil of the farmland. Microorganisms are important for healthy growth of the alfalfa plants. Explain why.

_____

_____ [2 marks]

f Many areas of farmland are created by deforestation. Give **one** reason why deforestation reduces the biodiversity of species in an area.

_____ [1 mark]

2. The camel is a mammal that is adapted to living in sandy deserts. Two physical adaptations of the camel are:

- large flat feet
- two rows of long eyelashes.

a Explain how each of these adaptations helps the camel to survive.

- Large flat feet: _____

- Two rows of long eyelashes: _____ [2 marks]

b Which **one** of the following is a living factor that could limit the size of a population of camels in the desert?

☐ light   ☐ competition   ☐ water   ☐ temperature   [1 mark]

3. The *Camelops* is a species of camel that lived many years ago. Scientists think that it was hunted to extinction.

a Which **two** of the following are also possible causes of extinction of a species?

☐ a change in climate   ☐ pollution

☐ improved medicines   ☐ breeding programmes   [2 marks]

b An increasing human population has caused many species to become close to extinction. Explain **one** way in which an increasing human population could cause extinction.

_____

_____ [2 marks]

4. Hena investigated the length of nettle leaves found in the shade and in the sunshine.

Her results are shown in the table.

Length of nettle leaf (cm)					
Nettles in the sunshine	4.3	3.9	4.4	4.7	5.1
Nettles in the shade	7.2	8.4	6.6	7.3	6.2

a Write a conclusion from Hena's results.

_____ [1 mark]

b Suggest an explanation for Hena's results.

_____

_____ [2 marks]

4.1 Using classification keys

Learning outcomes

- To use and construct keys to identify plants and animals
- To research the work of scientists studying the natural world

1. Complete the sentences by choosing words from this list.

| characteristics | height | invertebrates | kingdoms | species |

Living organisms can be classified according to their _____ .

Organisms are classified into large groups called _____ such as 'animal', 'plant' and 'fungi'.

[2 marks]

2. Draw lines to match each organism with the kingdom to which it belongs.

Organism

Crab

Moss

Mushroom

Kingdom

fungi

animal

plant

[2 marks]

3. Humans are part of the animal kingdom. Explain why humans are **not** classified as plants.

Worked Example

Humans are not classified as plants because human cells do not have cell walls (1). Also, human cells never contain chlorophyll and so they cannot make their own food like plant cells (1). [2 marks]

Remember

The plant kingdom contains multicellular organisms that contain chlorophyll and make their own food by photosynthesis.

4. Who invented the system of classification we use to classify organisms? Tick **one** box.

☐ **Charles Darwin**   ☐ **Francis Crick**

☐ **Carl Linnaeus**   ☐ **Robert Bakewell**

5. Why do scientists use classification keys?

_____ [1 mark]

6. Fatima collects leaves from four different types of tree. She is going to make an identification key so that each type of tree can be identified using its leaves.

Describe what Fatima needs to do to make an identification key.

Look carefully at the _____ .

Write _____ based on the characteristics. [2 marks]

7. Jalal finds some organisms in the school grounds. The diagrams below show the organisms he finds.

| stick insect | jumping spider | meal moth | hoopoe bird |

Jalal makes a table to record the characteristics of the organisms he finds.

Name of organism	Number of legs	Number of wings
Stick insect	6	0
Jumping spider		
Meal moth		
Hoopoe bird		

a Use the diagrams to complete the table. The first one has been done for you. [3 marks]

b Jalal uses the information in his table to construct a classification key. Complete the key by writing the missing question.

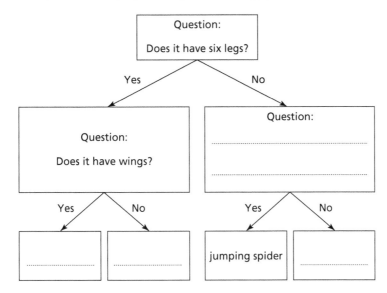

[1 mark]

43

c Write the name of each organism in the table in the correct box on the key above. One has been done for you. [2 marks]

8. Viruses cannot reproduce without invading a 'host' cell. Viruses are **not** put into any of the kingdoms. Explain why.

Challenge

_____

_____ [2 marks]

9. Aristotle was a Greek scientist who developed the first classification system. He divided all the known organisms into two groups – the plants and the animals.

Aristotle then divided each animal group into three smaller groups – land, water and air. Birds, bats and flying insects were grouped together as 'air animals'.

We now know that Aristotle's classification system does **not** work. It groups organisms that are completely different to one another in the same group.

Explain why Aristotle's classification of 'air animals' does **not** work.

_____

_____ [2 marks]

4.2 Genes and DNA

Learning outcomes
- To state what is meant by 'inheritance'
- To describe the roles of chromosomes and genes in inheritance
- To identify variation caused by genetic and environmental factors

1. Complete the sentences by choosing words from this list.

chromosomes	cytoplasm	enzymes	genes	nucleus

Information about physical characteristics is passed on from one generation to the

next by _____ , which are found on _____ in the

_____ of an egg cell and a sperm cell. [3 marks]

2. Which of the following **best** describes a gene? Tick **one** box only.

☐ A cell that carries genetic information

☐ A copy of all the DNA found in the body

☐ A damaged section of DNA

☐ A section of DNA that has instructions for a characteristic [1 mark]

3. Explain why a sperm cell and an egg cell only have half the number of chromosomes found in a normal body cell.

Show Me

The egg cell and sperm cell only contain half the number

of chromosomes because they _____

so that the fertilised egg has a _____ .

[2 marks]

> **Remember**
> Some people think DNA and genes are the same thing, but they are not. A gene is a section of DNA that contains information in the form of a code. The information may be used to produce a certain characteristic in an organism.

4. In the table below, put ticks in the correct columns to show whether the characteristic is:

- inherited only
- acquired only
- inherited and acquired.

The first one has been done for you.

Characteristic	Inherited only	Acquired only	Inherited and acquired
Skin colour			✓
Body mass			
A scar			
Shape of ear			

[3 marks]

5. The diagram shows two families. Some people in the diagram are able to roll their tongue.

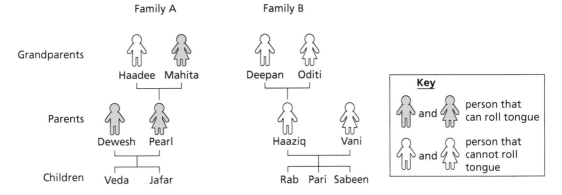

a Which **two** children are most likely to be able to roll their tongues?

Explain your answer.

_____

_____ [2 marks]

b Explain why Sabeen is **unlikely** to be able to roll her tongue.

_____ [1 mark]

6. Explain why identical twins look alike.

Challenge

_____

_____

_____ [2 marks]

7. In the 1940s, scientists knew that DNA was a very important molecule. However, they did not know about its structure.

In the 1950s, two scientists, Franklin and Wilkins, studied DNA using X-rays. Franklin and Wilkins were experts in a technique called X-ray diffraction.

Franklin produced an X-ray photograph that gave important clues about the structure of DNA. This allowed two other scientists, Watson and Crick, to produce a 3D model of DNA.

a What question were all the scientists trying to answer?

_____ [1 mark]

b Give **one** piece of evidence that Watson and Crick used to produce their model.

_____ [1 mark]

c Explain how the work of Watson and Crick was made possible by other scientists.

_____ [1 mark]

4.3 Selective breeding

Learning outcomes

- To describe how selective breeding can lead to new varieties and breeds
- To describe uses of selectively bred organisms

1. What is 'variation'? Tick **one** box.

☐ The differences in height between two species

☐ How colourful an organism is

☐ Differences between individuals

☐ The number of chromosomes in a species   [1 mark]

2. Give **two** examples of inherited variation seen in cats.

1. _____   2. _____   [2 marks]

3. Draw lines to match each organism with the correct reason for it being selectively bred.

Organism	Reason why it is selectively bred
cabbage	bred to supply the most milk or meat yields
flowers	bred to resist insect pests and diseases
buffalo	bred to give the most colour and fragrance

[2 marks]

4. Wheat plants have been selectively bred to produce varieties of wheat that produce a lot of seeds.

Show Me

Describe how a new variety of wheat plant is produced using selective breeding.

1. Choose individual plants that produce _____

2. _____

3. Select the offspring that produce

   _____ and breed them together.

Remember

Over many generations, selective breeding creates new varieties or breeds of a species with characteristics useful to humans.

4. Repeat until you have plants producing the desired number of seeds.   [3 marks]

5. Robert Bakewell is known for his work on selective breeding in the 1700s.

Challenge He selectively bred cattle to produce a breed that ate less food and put on more weight than any other breed.

Suggest why the process of selective breeding is still popular across the world today.

_____

_____ [2 marks]

4.4 Natural selection

Learning outcomes
- To state what is meant by 'evolution'
- To describe how organisms are adapted to where they live
- To explain Darwin's idea about how organisms evolve by natural selection

1. Complete the sentences by choosing words from this list.

variation	evolution	fossil	rocks	mutation

Darwin's theory of _____ explains how species have changed over

time. The theory is supported by evidence from observations of organisms and from

_____ records. [2 marks]

2. Draw lines to match the words with their correct definitions.

Word	Definition
Adaptation	the differences that exist between individuals
Variation	characteristic of an organism that allows it to survive in a certain habitat
Competition	when many organisms all need the same resources, which are in limited supply

[2 marks]

3. Charles Darwin caught different types of finches on the Galápagos Islands in the Pacific Ocean. He brought them back to London where they were analysed by John Gould. Each species had a different type of beak. The diagram shows two of these finches.

A
strong beak

B
long slender beak

Complete the table to show:

- the letter of the finch best suited to each type of food

- an explanation of how the beak is adapted to the food.

Finch	Main food source	Explanation
	Nectar from inside flowers	
	Seeds that need to be crushed	

[2 marks]

4. Explain what is meant by 'natural selection'.

Show Me

Organisms that are better _____

have a better chance of surviving and are therefore

more likely to _____ .   [2 marks]

Remember

Don't get *natural selection* and *evolution* mixed up. Natural selection is when a characteristic becomes more common in a population because organisms with the characteristic are more likely to survive. Evolution is the gradual changes of a species over time (which happens because of natural selection).

5. **Challenge** Describe **one** similarity and **one** difference between selective breeding and natural selection.

_____

_____

_____

[2 marks]

6. Lemmings are furry rodents naturally found in the Arctic tundra. In 1800 a population of lemmings was taken from the Arctic to an area with a warmer climate. The lemmings were left to live and reproduce in the new area. The population of lemmings was revisited 200 years later.

Some information about the lemmings is shown in the table.

Year	Average fur length of lemming (mm)
1800	12
2000	8

Explain how Darwin's theory of natural selection can explain this change.

_____

_____

_____

_____ [4 marks]

Self-assessment

Tick the column which best describes what you know and what you are able to do.

What you should know:	I don't understand this yet	I need more practice	I understand this
Organisms are classified into groups, the largest of which is a kingdom			
Linnaeus developed a classification system that is still used today			
Dichotomous keys may be used to identify plants and animals			
DNA is the genetic material that carries information determining how individual plants and animals develop			
Chromosomes are found in the nucleus of the cell and contain DNA			
Genes are sections of DNA that control the development of a specific characteristic			
Individuals inherit their DNA from their parents			
There is variation between organisms of the same species – this variation can be inherited, acquired or both			
Selective breeding can be used to develop new plant varieties and animal breeds, which have characteristics that make them more useful to humans.			
Selective breeding involves continuously choosing animals and plants with certain characteristics to breed together to produce new breeds/varieties with more desirable characteristics			

	I can't do this yet	I need more practice	I can do this by myself
Selective breeding has been happening for thousands of years and has led to the many varieties of certain species that we see today, including dogs, cats, cows and vegetables			
Evolution is the gradual change of living organisms over many generations			
Charles Darwin proposed the theory of evolution by means of natural selection			
Natural selection happens when there is variation in a species, with some organisms being (by chance) better adapted than others			
Natural selection means that those who are better adapted to their environment tend to be more likely to survive and have more offspring			

You should be able to:	I can't do this yet	I need more practice	I can do this by myself
Make observations and measurements			
Discuss and explain the importance of questions, evidence and explanations, using historical and contemporary examples			
Discuss the way that scientists work today and how they worked in the past, including reference to experimentation, evidence and creative thought			
Describe patterns seen in results			
Interpret results using scientific knowledge and understanding			

If you have ticked 'I don't understand this yet' or 'I can't do this yet' or mostly 'I need more practice', have another look at the relevant pages in the Student's Book. Then make sure you have completed all the questions in this Workbook chapter and the review questions in the Student's Book. If you have already completed all the questions ask your teacher for help and suggestions on how to progress.

Teacher's comments

Test-style questions

1. Surika finds some insects.

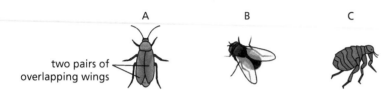

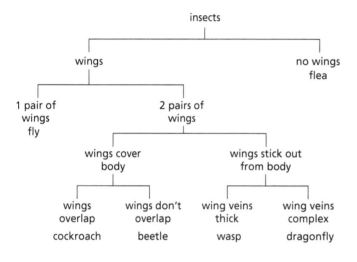

a   Use the key to identify **A**, **B** and **C**.

A: _____   **B**: _____

C: _____   [3 marks]

b   Surika's teacher says insects are classified as invertebrates. What is meant by 'invertebrate'?

_____   [1 mark]

2. Hardeep has brown eyes. His father also has brown eyes.

a   Describe how genetic information is passed on from the father to Hardeep.

_____

_____   [2 marks]

b Hardeep has an identical twin brother, Anand. Both have brown eyes, but Hardeep has a mass that is 6 kg more than Anand.

Explain why Hardeep and Anand's eye colour is identical but their masses are **not** identical.

_____

_____ [2 marks]

3. Many years ago, the ancestors of giraffes lived in Africa. Some had slightly longer necks than others. They fed on leaves in trees.

a Describe the variation in the giraffes that existed long ago.

_____ [1 mark]

b Explain why it could be an advantage for a giraffe to have a longer neck.

_____

_____ [2 marks]

c Giraffes in the African savannah today have longer necks than their ancestors.

Which process caused the changes in giraffes over the generations?

☐ **natural selection** ☐ **selective breeding** ☐ **growth** ☐ **variation** [1 mark]

d Which scientist is best known for his work on evolution? Tick **one** box.

☐ **James Watson** ☐ **Charles Darwin**

☐ **Robert Bakewell** ☐ **Gregor Mendel** [1 mark]

4. A farmer wants buffalo that produce a high milk yield. Describe how the farmer could use selective breeding to produce such animals.

_____

_____

_____

_____

_____ [3 marks]

Chemistry

Chapter 5: The Periodic Table

Chapter 6: Preparing salts

Chapter 7: Reactivity and rates of reaction

Chapter 8: Energy transfers in chemistry

5.1 Structure of an atom

Learning outcomes

- To describe the structure of an atom
- To learn about the methods and discoveries of Rutherford
- To think about how scientists build on and improve earlier ideas by asking questions, designing experiments and then testing them
- To compare the structures of some simple atoms

1. Atoms have no overall charge – they are neutral. Which one of the sentences below explains why?

Tick **one** box.

☐ They contain the same number of electrons and protons.

☐ They contain the same number of protons and neutrons.

☐ They contain neutrons, which have no charge.

☐ The nucleus contains the same number of positive and negative particles.   [1 mark]

2. The nucleus of an atom contains 3 protons and 4 neutrons. Which statement about the atom is true?

Tick **one** box.

☐ It has an atomic number of 4

☐ It has an atomic number of 3

☐ It has three electron shells

☐ It has an atomic number of 7   [1 mark]

3. This is a diagram of a lithium atom.

Label the diagram by choosing words from this list.

| period   group   proton   neutron |
| proton number   electron |
| nucleus   electron shell (orbit) |

[5 marks]

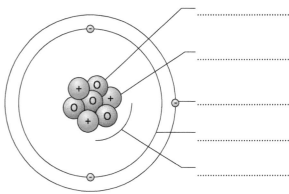

4. The following diagram shows the 'plum pudding' model of the atom. It was first suggested by the scientist J.J. Thomson in 1904.

Complete these sentences about Thompson's model using words from this list.

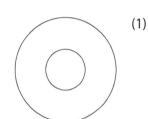

negatively	neutrally	positively	electrons
	protons	neutrons	nucleus

The atom is a _____ charged ball with negatively charged

_____ in it. In 1912, the idea of having a _____

in the centre of the atom was not the accepted model.   [3 marks]

5. The symbol for fluorine in the Periodic Table is $^{19}_{9}F$.   (1)

 Worked Example

Complete the diagram of the fluorine atom by adding the protons, neutrons and electrons.

Proton number = number of protons = number of electrons = 9 (1)

Number of neutrons = mass number – proton number = 19 – 9 = 10 (1)

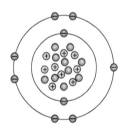

[2 marks]

Remember
You need to use the numbers given in the symbol to work out the number of protons, neutrons and electrons in the atom.

6. The symbol for carbon in the Periodic Table is $^{12}_{6}C$. Describe the electronic structure of a carbon atom.

 Show Me

There are _____ protons in the nucleus.

There are 6 electrons orbiting the nucleus in

_____ . There are 2 electrons in the first

shell and _____ .

There are _____ neutrons in the nucleus.

[4 marks]

Remember
Use the information provided in the question or look it up in the Periodic Table. You are not expected to remember the atomic structure of each atom but you do need to know how to use the information to work it out.

7. These diagrams show two different models of the atom.

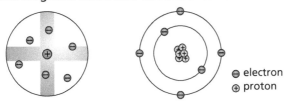

Ernest Rutherford's model of the atom (1911)   Niels Bohr's model of the atom (1932)

⊖ electron
⊕ proton

State **two** similarities and **one** difference between the two models.

_____

_____

_____

_____ [3 marks]

8. Sodium has atomic number 11; chlorine has atomic number 17.

Challenge Compare the structure of a sodium atom with that of a chlorine atom.

_____

_____

_____

_____

_____ [4 marks]

9. **Modelling the atom**

The model of the atom has changed throughout time. This is because of the creative thought of scientists and the experimental evidence they collected.

The diagrams show five different models of atoms suggested by the scientists listed in the box below.

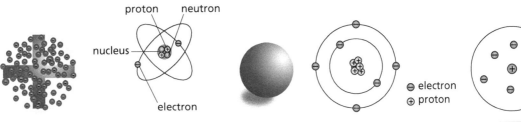

proton   neutron

nucleus

electron

⊖ electron
⊕ proton

| John Dalton, 1803 | Joseph John Thomson 1897 | Ernest Rutherford, 1911 |
| Niels Bohr, 1913 | James Chadwick, 1932 | |

On a large piece of paper, draw this table.

Scientist	Date	Model of atom	Main ideas shown by model
John Dalton	1803		Atoms are like tiny hard spheres

a   Use the information provided to complete the table – the first row has been done for you.

If you have access to the internet you may want to find out more by looking up the scientists.

[2 marks for each correct row]

b   Underneath your table, write a paragraph to describe how the model of the atom has changed over time.   [5 marks]

5.2 Trends in the Periodic Table

Learning outcomes

- To describe trends in periods and groups in the Periodic Table
- To explain how the Periodic Table was developed

1. Neon is in Group 8 of the Periodic Table. Argon is below neon in the group.

Which of these statements is true? Tick **one** statement.

☐ Neon is more reactive than argon.

☐ Neon is less reactive than argon.

☐ Neon is an inert gas.

☐ Neon is a solid. [1 mark]

2. The diagram shows the outline of the Periodic Table.

a Shade a group of non-metals blue. [1 mark]

b Shade the Group 2 metals green. [1 mark]

c Draw a red box round the Period 2 elements. [1 mark]

d Which of the three elements indicated **X**, **Y** and **Z** have atoms with one electron in their outer shells?

_____ [1 mark]

e Which of the three elements have atoms with the most electron shells?

_____ [1 mark]

3. Complete the sentences using words from this list.

| faster | slower | carbon dioxide | oxygen | hydrogen | above | below | next to |

When lithium reacts with water, _____ gas is formed. Sodium is

_____ lithium in their group. The rate of reaction of sodium with

water is _____ than the rate of reaction of lithium and water. [3 marks]

4. Read the following information about three Group 7 elements.

Show Me

At room temperature:

- bromine is a brown liquid

- iodine is a purple-black solid

- chlorine is a green gas.

Write down in order of increasing atomic number these three Group 7 elements.
Give a reason for your answer.

Going down the group the mass of the atoms/molecules

_____ so the melting and boiling points

also _____ . At room temperature, a

gas has a _____ boiling point than a

liquid, and a solid has a _____ melting

point than a liquid. Therefore, going down the group, the order of the elements is

_____ . [4 marks]

Remember
Use information that is given in questions to help you to give reasons for your answer. You need to be able to link information to what you already know.

5. Magnesium and calcium are both in Group 2 of the Periodic Table. Magnesium is above calcium. When magnesium is added to hydrochloric acid you can see bubbles of hydrogen gas.

Predict what you will observe when calcium is added to some hydrochloric acid.

_____

_____ [2 marks]

6.

Challenge

Explain why elements in the same group of the Periodic Table have similar chemical properties.

_____

_____ [2 marks]

7.

This is Mendeleev's Periodic Table from 1869.

I							
H 1.01	II	III	IV	V	VI	VII	
Li 6.94	**Be** 9.01	**B** 10.8	**C** 12.0	**N** 14.0	**O** 16.0	**F** 19.0	
Na 23.0	**Mg** 24.3	**Al** 27.0	**Si** 28.1	**P** 31.0	**S** 32.1	**Cl** 35.5	VIII
K 39.1	**Ca** 40.1		**Ti** 47.9	**V** 50.9	**Cr** 52.0	**Mn** 54.9	**Fe** 55.9 **Co** 58.9 **Ni** 58.7
Cu 63.5	**Zn** 65.4			**As** 74.9	**Se** 79.0	**Br** 79.9	
Rb 85.5	**Sr** 87.6	**Y** 88.9	**Zr** 91.2	**Nb** 92.9	**Mo** 95.9		**Ru** 101 **Rh** 103 **Pd** 106
Ag 108	**Cd** 112	**In** 115	**Sn** 119	**Sb** 122	**Te** 128	**I** 127	
Ce 133	**Ba** 137	**La** 139		**Ta** 181	**W** 184		**Os** 194 **Ir** 192 **Pt** 195
Au 197	**Hg** 201	**Ti** 204	**Pb** 207	**Bi** 209			
			Th 232		**U** 238		

Explain why Mendeleev left gaps in his version of the Periodic Table.

_____

_____ [2 marks]

8.

Practical

Jamila and Yuri are investigating the Periodic Table. They find some data on the internet about a group of non-metals. The data is shown in the table below.

Element						
Atomic number	2	10	18	36	54	86
Density in g/dm³	0.18	0.90	1.78	3.71	5.85	9.97

a Complete the first row of the table by writing in the names of the elements. To do this you will need to look at a copy of the Periodic Table. [3 marks]

b Jamila thinks that there is a pattern in the data. Plot the data on the graph axes below to find out if she is right. Add a line of best fit.

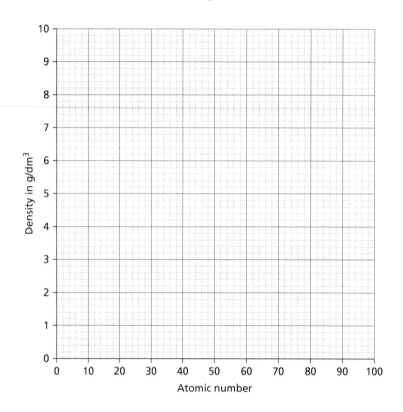

[4 marks]

c Describe the pattern between atomic number and density using the information in your graph.

_____

_____ [2 marks]

d Use the graph to predict the density of the elements with atomic number:

   i 31 (gallium) _____ [1 mark]

   ii 49 (indium) _____ [1 mark]

e Yuri thinks that the predictions for the density of gallium and indium are wrong.

Jamila disagrees because she thinks the graph is accurate.

Explain why Yuri is correct to disagree.

_____

_____ [2 marks]

Self-assessment

Tick the column which best describes what you know and what you are able to do.

What you should know:	I don't understand this yet	I need more practice	I understand this
Models in science are based on evidence collected in scientific experiments			
Rutherford concluded from his gold foil experiment that an atom has a small, central, positively charged nucleus, but most of it is empty space			
The model of the atom we currently use has a nucleus with electrons moving round it in shells			
The nucleus contains positively charged protons and neutrons, which have no charge			
Elements are arranged in order of atomic number			
Elements in the Periodic Table are arranged in columns called groups, and rows called periods			
Elements in the same group share similar properties but there are trends, such as reactivity and boiling point, within the groups			
The Periodic Table was developed over many years based on ideas and evidence from different scientists			

You should be able to:	I can't do this yet	I need more practice	I can do this by myself
Discuss and explain the importance of questions, evidence and explanations, using historical and contemporary examples			
Discuss the way that scientists work today and how they worked in the past, including reference to experimentation, evidence and creative thought			

If you have ticked 'I don't understand this yet' or 'I can't do this yet' or mostly 'I need more practice', have another look at the relevant pages in the Student's Book. Then make sure you have completed all the questions in this workbook chapter and the review questions have another look at the relevant pages in the Student's Book. If you have already completed all the questions ask your teacher for help and suggestions on how to progress.

Test-style questions

1. Look at the diagram of a nitrogen atom.

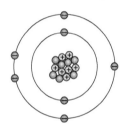

a How many electrons are in the atom? _____ [1 mark]

b How many neutrons are in the atom? _____ [1 mark]

c How many protons are in the atom? _____ [1 mark]

d Write down the chemical symbol of nitrogen _____ [1 mark]

e The element phosphorus is below nitrogen in the Periodic Table. Phosphorus has an atomic number of 15. Describe **two** ways in which the structure of a phosphorus atom is different from a nitrogen atom.

_____

_____ [2 marks]

2. Look at the information about some different atoms:

helium   fluorine   sodium   aluminium

$_2^4He$ $_9^{19}F$ $_{11}^{23}Na$ $_{13}^{27}Al$

a Which atom has the same number of protons and neutrons?

_____ [1 mark]

b Which atom is in Group 1 of the Periodic Table? _____ [1 mark]

c Which atom has seven electrons in its outer shell? _____ [1 mark]

d Which atom has the most neutrons in its nucleus? _____ [1 mark]

3. Today the Periodic Table is based on the groups that Dmitri Mendeleev suggested in 1869.

Complete these sentences about his work.

a Mendeleev arranged the elements in order of increasing _____ . [1 mark]

b Mendeleev observed that some elements have similar _____

and _____ properties. [1 mark]

c Mendeleev left gaps in his Periodic Table so that new _____

_____ . [1 mark]

d Why was Mendeleev's way of grouping elements so important?

_____

_____ [2 marks]

4. The diagram represents the Periodic Table of elements.

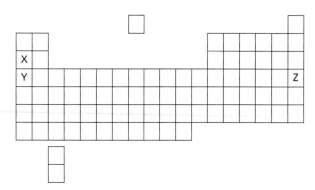

X, **Y** and **Z** represent elements in different positions in the Periodic Table.

a Which letter represents the element with the highest atomic number?

_____ [1 mark]

b Which letter represents the element which is a gas at room temperature?

_____ [1 mark]

c Elements **X** and **Y** react with cold water to produce hydrogen gas.

  i Describe what you would observe when **X** and **Y** are added separately to cold water.

  _____ [1 mark]

  ii Explain which element will react with water fastest. Give a reason for your answer.

  _____

  _____ [1 mark]

  iii Explain why elements **X** and **Y** have similar chemical properties.

  _____

  _____ [2 marks]

6.1 Reactions with acids

Learning outcomes

- To describe the reactions of acids with metals and metal carbonates
- To write and balance symbol equations to describe reactions

. . . .

1. The formula of calcium sulfate is $CaSO_4$. What is the total number of atoms present?

Tick **one** box.

☐ 1 ☐ 3 ☐ 6 ☐ 12 [1 mark]

2. Name the salt made when magnesium reacts with hydrochloric acid.

Tick **one** box.

☐ magnesium chloride ☐ magnesium hydrochloric

☐ hydrogen gas ☐ magnesium sulfate [1 mark]

3. The diagram shows some sulfuric acid in a measuring cylinder.

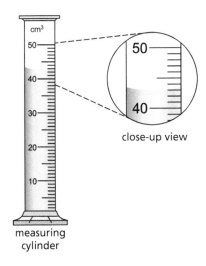

close-up view

measuring
cylinder

How much sulfuric acid is in the measuring cylinder?

_____ [2 marks]

The final stage in making a salt is crystallisation.
Explain what 'crystallisation' means.

Crystallisation is a separation technique that is used
to separate a solid from a solution (1). The liquid
evaporates leaving crystals of the solid
behind. (1)   [2 marks]

Remember
If there are two marks for
a written question like this
then you must make sure
your answer has enough
detail to get both marks.

5.

Read the safety information about calcium.

Show Me

Practical

- Calcium reacts readily with water (or acids) to produce hydrogen, an extremely flammable gas.
- Contact with moisture forms calcium oxide or calcium hydroxide which can irritate the eyes and skin.
- It burns vigorously, but is difficult to ignite.

Describe the safety precautions you should take when making a salt using calcium metal.

Keep it away from naked flames because one of

the substances involved is _____ .

Wear _____ to protect your eyes

from splashes, which are _____ .

Do not touch calcium with your hands – wear _____ .   [4 marks]

Remember
Read questions carefully
to find the main points in
the text.

6.

Practical

Dilute hydrochloric acid is classified as 'low hazard'. Describe a safety precaution you should always take when using dilute acids.

_____   [1 mark]

7.

Challenge

Explain why some salts have large crystals, while others have small crystals.

_____

_____

_____   [2 marks]

8. Making magnesium sulfate

Oliver and Gabriella want to make some magnesium sulfate. They have been given the following method.

- Add excess metal to 25 cm³ of dilute acid until no more dissolves.

- Filter off the excess metal.

- Evaporate until some solid appears.

- Leave to cool.

- Filter.

a Name the starting materials they should use.

_____ [2 marks]

b The diagram shows four pieces of apparatus that can be used to measure volumes of liquids.

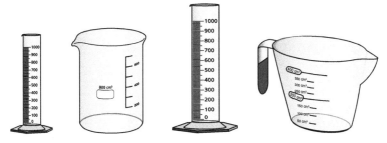

Name the piece of apparatus Oliver and Gabriella should use to measure out the acid. Give a reason for your choice.

_____

_____ [2 marks]

c What does 'excess metal' mean?

_____ [1 mark]

d Gabriella wants to see the crystals. She wants to continue heating when the solid starts to appear, but Oliver disagrees. He says to get good crystals you must leave the solution to cool. Why is Oliver correct?

_____

_____ [2 marks]

e Oliver is about to pour the mixture into the filter funnel when Gabriella stops him because he has forgotten the filter paper.

    i Where should the filter paper be?

    _____ [1 mark]

    ii Explain why the mixture needs to be filtered.

    _____ [2 marks]

f Write a word equation for the reaction.

_____ [2 marks]

6.2 Methods for making salts

Learning outcomes
- To describe how to make salts in the laboratory
- To decide what apparatus and reactants to use to make a particular salt

1. Which word equation describes the reaction between an acid and a carbonate?

Tick **one** box.

☐ acid + carbonate → salt + water

☐ acid + carbonate → salt + hydrogen

☐ acid + carbonate → salt + water + carbon dioxide

☐ acid + carbonate → salt + water + hydrogen   [1 mark]

2. **Making a salt starting with two solutions**

Practical   Blessy and Mia want to make some sodium chloride crystals. The starting materials available are sodium carbonate solution and dilute hydrochloric acid.

a Write a word equation for the reaction.

_____ [2 marks]

b Before starting the practical work the girls do a risk assessment. Dilute sodium carbonate solution and dilute hydrochloric acid are both classified as 'low hazard'.

Give **one** safety precaution the girls should take.

_____ [1 mark]

c Blessy and Mia are not sure how to carry out the experiment so they do a trial experiment.

- Mia measures out 25 cm³ of hydrochloric acid and pours it into a beaker.

- Blessy then adds 25 cm³ of sodium carbonate solution and observes the reaction.

- They evaporate the mixture until some solid appears before leaving it to cool.

  i What does Blessy observe?

  _____ [1 mark]

  ii How do they know when the reaction is over?

  _____ [1 mark]

d Mia is not happy with this method. She is worried that there might be some unreacted acid in the beaker. Describe a test she could carry out to see if she is right.

_____

_____ [2 marks]

e Mia carries out the test and finds that she is right. Describe how the girls should change their method to avoid leaving some unreacted acid at the end of the reaction.

_____

_____

_____

_____ [2 marks]

f Blessy is now concerned that the sodium chloride crystals will be impure. So she:

- adds some charcoal

- filters the mixture

- evaporates off the water

- leaves the crystals to form as the solution cools.

    i Why does Blessy add charcoal?

_____ [1 mark]

    ii Why does Blessy filter the mixture?

_____ [1 mark]

3. Describe how the pH changes as sodium carbonate is added to some sulfuric acid.

Show Me

At the start of the reaction the pH is

_____ because there is only sulfuric

acid in the beaker. When all the sulfuric acid has

reacted, the pH is _____ because the

_____ and water are neutral. If more

sodium carbonate is added the pH will _____

until it reaches 10 or 11. [4 marks]

Remember
Break down the problem into small parts or stages. Then describe each stage in turn.

4. What would you expect the pH of copper chloride solution to be? Give a reason for your answer.

_____

_____ [2 marks]

5. Hassan was making potassium sulfate by adding potassium carbonate powder to some sulfuric acid. During the final stages of the reaction, the potassium carbonate did not fizz when it was added. When Hassan measured the pH he found it was 10.5.

Challenge

Explain his result.

_____

_____

_____

_____ [3 marks]

Self-assessment

Tick the column which best describes what you know and what you are able to do.

What you should know:	I don't understand this yet	I need more practice	I understand this
Symbol equations show the chemical symbols and formulae of the reactants and products – they also show the numbers of atoms in the reactants and products			
Metals and metal carbonates react with acids to produce salts			
The type of salt made depends on the acid used			
Salts can be made by reacting metals or insoluble metal carbonates with an acid			
The metal or metal carbonate is added in excess so all the acid reacts			
The excess metal or metal carbonate is removed using filtration. The water is evaporated from the salt solution to form dry salt crystals			
Salts can also be made by reacting soluble metal carbonates (in solution) with an acid. An indicator is used to show when all the acid has reacted. The same procedure is used to crystallise the salt that has been made			

You should be able to:	I can't do this yet	I need more practice	I can do this by myself
Decide which measurements and observations are necessary and what apparatus to use			
Decide which apparatus to use and assess any hazards in the laboratory, field or workplace			
Make sufficient observations and measurements to reduce error and make results more reliable			
Explain results using scientific knowledge and understanding – communicate this clearly to others			
Use a range of materials, apparatus and control risks			

If you have ticked 'I don't understand this yet' or 'I can't do this yet' or mostly 'I need more practice', have another look at the relevant pages in the Student's Book. Then make sure you have completed all the questions in this Workbook chapter and the review questions in the Student's Book. If you have already completed all the questions ask your teacher for help and suggestions on how to progress.

Test-style questions

· ·

1. Salts can be prepared by the reaction of acids with metals.

a **i** Complete this equation:

acid + metal → salt + _____   [1 mark]

ii Circle the word that best describes the reaction.

oxidation   **combustion**   **neutralisation**   **displacement**   [1 mark]

b Sodium sulfate is an important salt used to make detergents.

The table lists some substances. Tick the name of the acid used to make sodium sulfate.

Hydrochloric acid	
Sulfuric acid	
Nitric acid	

[1 mark]

c Sodium sulfate can be made by the reaction of an acid and carbonate.

Complete the word equation for this reaction.

_____ carbonate + _____ acid →

sodium sulfate + _____ + _____   [2 marks]

74

2. Youssef is making zinc chloride. He uses this method:

Practical

- add one spatula of zinc to some hydrochloric acid and stir
- keep adding zinc and stirring until there is an excess
- remove the excess zinc
- crystallise the salt solution.

a State **one** hazard when making the salt.

_____ [1 mark]

b Describe **one** way of reducing the risk of harm.

_____ [1 mark]

c When will Youssef know it is time to stop adding the zinc?

_____ [1 mark]

d On a separate piece of paper, draw a labelled diagram to show how the excess zinc is removed. [5 marks]

e Write a word equation for the reaction.

_____ [2 marks]

3. Acids and bases are commonly found around the home.

Practical

a Some indigestion tablets contain calcium carbonate which neutralises excess stomach acid (hydrochloric acid).

i Complete the word equation for the reaction.

calcium carbonate + hydrochloric acid → _____ +

_____ + _____ [1 mark]

ii How does the pH change in the stomach after taking these tablets?

_____ [1 mark]

b Ammonium sulfate is a salt that is used to make fertilisers. Some people use ammonium sulfate in their garden.

Safia and Lily were making ammonium sulfate in the laboratory. Here is their method:

- pour 100 cm³ of sulfuric acid into a beaker

- add some indicator

- add some ammonium hydroxide until the solution is neutral

- add some charcoal

- filter

- evaporate and cool

- filter.

   i Name the piece of apparatus the girls should use to measure out the sulfuric acid.

   _____ [1 mark]

   ii Explain why Safia and Lily added some indicator to the acid.

   _____ [1 mark]

   iii Why did the girls filter the mixture twice?

   _____

   _____ [2 marks]

   iv Describe how Lily and Safia evaporated the final solution. You can include a labelled diagram as part of your answer if it helps.

   _____

   _____

   _____

   _____ [3 marks]

7.1 The reactivity series

Learning outcomes

- To understand that some metals are more reactive than others
- To use the reactivity series to compare the reactivity of metals
- To make enough observations and measurements in an investigation to reduce error and increase the reliability of results

1. Look at the diagram. It shows some metals reacting in water.

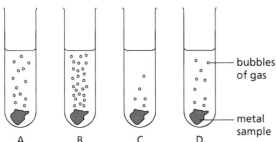

A   B   C   D   bubbles of gas   metal sample

Put the metals in order of reactivity, starting with the most reactive.

Tick **one** box.

☐ A B C D   ☐ A D C B   ☐ B A D C   ☐ D B C A   [1 mark]

2. Which of the following metals shows the fastest reaction with cold water?

Tick **one** box.

☐ Copper   ☐ Sodium   ☐ Zinc   ☐ Magnesium   [1 mark]

3. Choose words from the list to complete the sentences that follow.

| reactive | gold | silver | unreactive | oil | water | acid | middle | bottom | top |

The Californian gold rush began in 1848 when _____ was found by

James Marshall at Sutter's Mill. Gold is very _____. It does not react with

_____ or oxygen, which is why it is sometimes found in river beds. Gold is

found towards the _____ of the reactivity series of metals.   [4 marks]

4.

A teacher carefully cut some samples of metals and left them in the air. The class timed how long it took for the cut edge of the metals to go dull and recorded the results on scrap paper. The times were:

Practical

lithium: 92 s   potassium: 13 s   calcium: 2 min 30 s   sodium 57 s.

a Put the data into this blank table. Start by giving each column a heading.

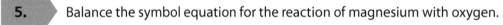

[3 marks]

b Underline the metal showing the fastest reaction. [1 mark]

c Draw a circle round the slowest reaction. [1 mark]

5. Balance the symbol equation for the reaction of magnesium with oxygen.

Worked Example

___ Mg + ___ $O_2 \rightarrow$ ___ MgO

Draw the atoms on both sides of the equation.

To balance the oxygen atoms multiply all the atoms in MgO by 2. (1)

To balance the magnesium atoms, multiply the magnesium atoms on the left by 2. (1)

Remember

To balance a symbol equation you must have the same number of atoms on each side of the equation. You cannot change the number of atoms in a formula.

The numbers of atoms on each side of the equation are the same.

Write the multipliers in front of the atoms or compounds:

$2Mg + O_2 \rightarrow 2MgO$

[2 marks]

6. Complete the symbol equation for the reaction of sodium with water.

Show Me

_____Na + _____$H_2O \rightarrow$ _____ NaOH + _____ H_2

Draw the atoms on each side of the equation:

Start by trying to balance the hydrogen atoms. Multiply all the atoms in

_____ by 2.

To balance the oxygen atoms, multiply all the atoms in _____

by 2.

To balance the sodium atoms, multiply the sodium atoms on the

_____ by 2.

The numbers of atoms on each side of the equation are the same.

Go back to the equation at the start of the question and write in the multipliers in front of the atoms and compounds.   [3 marks]

7. When calcium reacts with water, the reaction is:

- more vigorous (faster) than the reaction between zinc and water

- less vigorous (slower) than the reaction between magnesium and water.

Predict how the metals will react in hydrochloric acid by drawing bubbles in the test tubes.

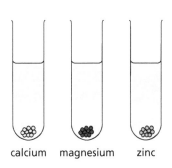

calcium   magnesium   zinc

[1 mark]

8.

Aluminium is above iron in the reactivity series of metals.

Observations show that objects made from aluminium do not corrode as quickly as those made from iron. Use your research to explain why aluminium corrodes less quickly than expected to explain the observations.

_____

_____

_____ [3 marks]

9.

Look at the diagram of the reactivity series – it shows the metals in order of reactivity.

Read these statements:

- Copper beads over 10 000 years old have been found in Northern Iraq.

- The ancient Hittites of Asia Minor extracted iron in the 1500s BC.

- Sodium metal was first extracted in 1807.

Most reactive	K	potassium
	Na	sodium
	Ca	calcium
	Mg	magnesium
	Al	aluminium
	C	carbon
	Zn	zinc
	Fe	iron
	Sn	tin
	Pb	lead
	H	hydrogen
	Cu	copper
	Ag	silver
Least reactive	Au	gold
	Pt	platinum

a Describe the pattern you can see between a metal's place in the reactivity series and its discovery.

_____ [1 mark]

b Suggest an explanation for the pattern.

_____

_____

_____

_____

_____ [3 marks]

10. **Investigating the reactivity of metals**

Priya and Carlos are investigating the reactivity of metals with different liquids.

Experiment 1: They add small pieces of the metals to water.

Experiment 2: They add small pieces of the metals to dilute hydrochloric acid.

a What two variables should Priya and Carlos control?

_____

_____ [2 marks]

b Describe the two safety precautions they should take.

_____

_____ [2 marks]

c Here are the results of the experiments.

Metal	Experiment 1 Reaction with water	Experiment 2 Reaction with hydrochloric acid
A	Lots of bubbles	Lots of bubbles forming very quickly
B	No reaction	Some bubbles on surface of metal
C	No reaction	No reaction
D	A few bubbles on surface of metal	Lots of bubbles

Write the metals in order of reactivity, starting with the most reactive.

_____ [1 mark]

d State which metal could be copper. Give a reason for your answer.

_____

_____ [2 marks]

e State which metal could be calcium. Give a reason for your answer.

_____

_____ [2 marks]

f Write a word equation for the reaction of calcium with water.

_____ [2 marks]

g Balance the symbol equation for the reaction of calcium with hydrochloric acid.

$$Ca + \underline{\hspace{2cm}} HCl \rightarrow CaCl_2 + H_2$$ [1 mark]

h The students decide to repeat the experiment using dilute sulfuric acid.

Predict what will happen when some sulfuric acid is added to metal **A** and to metal **C**.

_____

_____   [2 marks]

7.2 Displacement reactions

Learning outcomes

• To give examples of displacement reactions
• To use the reactivity series to predict whether or not a displacement reaction will occur

1. Complete the sentences that follow by choosing from these words:

| more | less | oxidation | neutralisation | displacement |
| non-metal | metal | mixtures | compounds |

A _____ reaction occurs when a _____ reactive metal

displaces a less reactive _____ from one of its _____ .

[4 marks]

2. The diagram shows the reactivity series of metals.

a Which metal can be used to displace magnesium from magnesium nitrate?

Tick **one** box.

☐ iron   ☐ copper

☐ sodium   ☐ aluminium   [1 mark]

Most reactive
K   potassium
Na sodium
Ca calcium
Mg magnesium
Al aluminium
C   carbon
Zn zinc
Fe iron
Sn tin
Pb lead
H   hydrogen
Cu copper
Ag silver
Au gold
Least reactive
Pt platinum

b Which metal can be used to displace iron from iron sulfate?

Tick **one** box.

☐ **iron**   ☐ **magnesium**   ☐ **lead**   ☐ **copper**   [1 mark]

3. The first diagram shows an iron nail being put into some copper sulfate solution. The second diagram shows what it looks like after 20 minutes.

Show Me

a Look at the diagrams and write down your observations.

During the chemical reaction the blue

copper sulfate solution has become

_____. Part of the silver-coloured

nail has become covered with _____ .   [2 marks]

Remember
An observation is what you *see happening*. You do not need to explain observations.

b A displacement reaction has taken place. Complete the word equation.

iron + copper sulfate → _____ + _____   [2 marks]

4. This diagram shows two metals dipped in copper sulfate solution.

How could you tell if zinc or iron is more reactive from this experiment?

_____

_____

[1 mark]

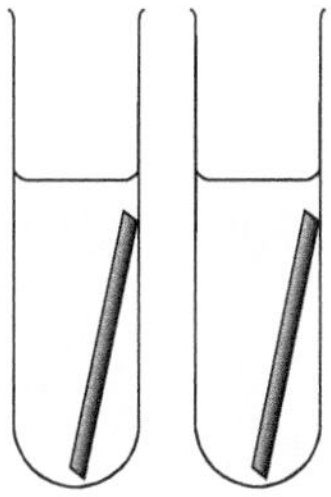

zinc   iron

5.

Look at the displacement reactions shown in the two word equations below.

barium + nickel chloride → barium chloride + nickel

nickel + tin sulfate → nickel sulfate + tin

Put the metals in order of reactivity with the most reactive metal first.

Explain your answer.

_____

_____

_____ [3 marks]

Predicting reactivity

6.

Pierre and Ravjiv are investigating the reactivity of different metals using displacement reactions. This is their method:

- Measure out 10 cm³ of four different metal sulfate solutions into test tubes.

- Add pieces of different metals to the test tubes.

- Carefully observe what happens.

- Record the results.

The results are listed in the table.

	Magnesium sulfate	Zinc sulfate	Iron sulfate	Copper sulfate
Zinc	no reaction		reaction	reaction
Magnesium		reaction	reaction	reaction
Unknown	no reaction	reaction	reaction	reaction
Iron	no reaction	no reaction		reaction
Copper	no reaction	no reaction	no reaction	

a Suggest why the boys shaded some parts of the table grey.

_____ [1 mark]

b Put the metals in order of reactivity, starting with the most reactive.

_____ [2 marks]

c Use your knowledge of the reactivity series to name the unknown metal. Give a reason for your answer.

_____

_____ [2 marks]

7.3 Rates of reaction

Learning outcomes

- To define the term 'rate of reaction'
- To state the different factors that affect how fast a chemical reaction happens
- To explain how these factors affect the rate of reaction
- To plan and carry out investigations into the factors that affect rate of reaction

1. Savita is cooking the dinner. She has forgotten to boil the potatoes. She wants them to cook fast so they are ready on time. Which potatoes will cook the fastest?

Tick **one** box.

☐ Whole potatoes

☐ Potatoes cut in half

☐ Potatoes cut in quarters

☐ Potatoes chopped into lots of small pieces [1 mark]

2. Petro is making some biscuits. The first step is to melt some butter in a pan. What could he do to make the butter melt quicker?

Tick **one** box.

☐ Put the pan in the fridge

☐ Heat the pan

☐ Spread the butter on the bottom of the pan

☐ Add some water [1 mark]

3. Draw three lines between the reacting particles and their concentration.

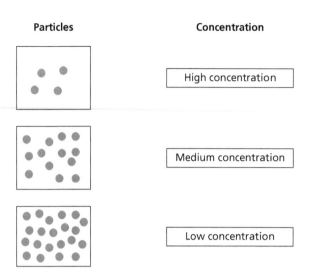

Particles

Concentration

High concentration

Medium concentration

Low concentration

[2 marks]

4. Complete the sentences that follow by choosing words from the box.

catalyst	slower	faster	particle	small	collision	rate	large

A _____ is a substance that increases the _____ of

reaction but is not used up during the reaction.

Small pieces of solid, especially powders, have a _____ surface area.

The larger the surface area, the _____ the reaction. [4 marks]

5. Marcus was investigating the rate of reaction between calcium carbonate and hydrochloric acid. He changed the particle size of the calcium carbonate and kept all other variables the same. Marcus timed how long it took for the reaction to take place.

Draw lines to join each particle size to the appropriate reaction time.

Particle size

| Single, large chip |
| Powder |
| Small chips |
| Medium chips |

Reaction time

| 1.0 minute |
| 4.2 minutes |
| 8.4 minutes |
| 6.5 minutes |

[3 marks]

6. Use ideas about particle theory to explain why increasing the temperature of a solution increases the rate of reaction.

Worked Example

For a reaction to occur particles must collide. (1)

Particles in a hot solution have more kinetic energy so they move around more quickly. (1)

This means that they will collide with other particles more frequently. (1) [3 marks]

Remember
When a question asks you to 'use ideas about' you must think about what you have learned and then apply it to the question.

7. Use ideas about particle theory to explain why increasing the concentration of a solution increases the rate of reaction.

Show Me

For a reaction to occur particles must _____ .

Increasing the concentration means the

particles are _____ .

This means that particles will collide with other

particles _____ . [3 marks]

Remember
'Explain' means that you should say what happens and why it happens.

8. The diagram shows three glasses of orange squash.

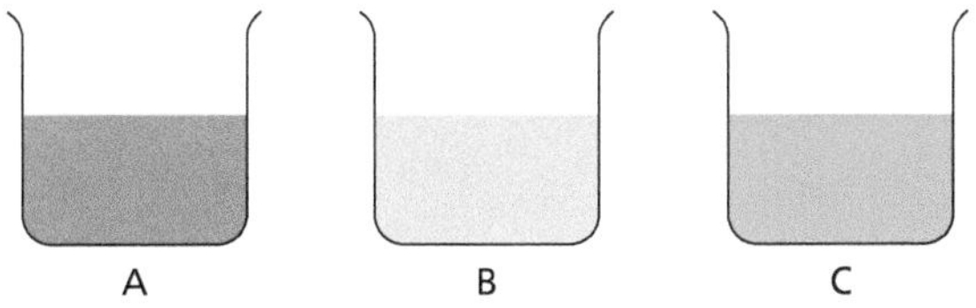

A   B   C

Which glass contains the lowest concentration of orange squash? Give a reason for your answer.

_____

_____

_____ [3 marks]

9. Mike and Chen carried out an investigation to see if the concentration of an acid has any effect on the time taken for a piece of magnesium ribbon to completely react.

This table shows their results.

Concentration of acid (%)	Time taken to dissolve (s)
100	13
75	22
60	35
50	50
38	100
30	145
25	250

a Plot the points on the axes below and draw a line of best fit.

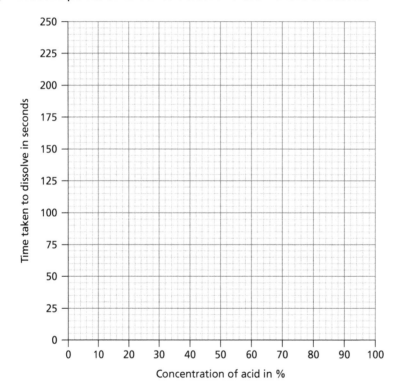

Concentration of acid in %

[4 marks]

b Use your graph to describe how increasing the concentration of the acid affects the rate of reaction.

_____ [1 mark]

c Use your graph to predict the time taken for the reaction if the concentration of the acid was 90%.

_____ [1 mark]

d Use your knowledge of the particle theory to explain your answer to part (b).

_____

_____ [3 marks]

10.

Challenge

Angelique is making a camp fire with her little brother. She asks him to go and find some small pieces of wood to burn. Her brother returns with a large tree trunk.

Explain why Angelique sends her brother away again saying 'Only small pieces of wood will work'.

_____

_____

_____ [3 marks]

11. The contact process in an industrial method for making sulfuric acid that uses a vanadium oxide catalyst. Write **two** reasons why many industrial processes use a catalyst.

_____

_____ [2 marks]

12.

Practical

Decomposing hydrogen peroxide

Hydrogen peroxide can be decomposed into water and oxygen.

Anatasia and Ahmed are doing an investigation to find the best catalyst for the reaction.

This is their method:

- Set up the apparatus shown in the diagram.
- Add 25 cm³ of hydrogen peroxide to each cylinder at the same time.
- Measure the height of the bubbles in the cylinder when bubbling stops.

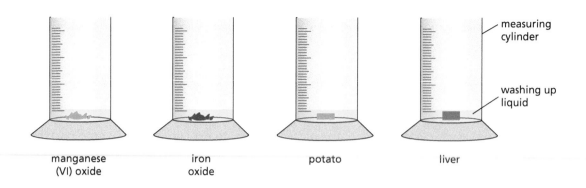

manganese (VI) oxide   iron oxide   potato   liver

a   What is a catalyst?

_____

_____ [2 marks]

b   Write a word equation for the reaction.

_____ → _____ + _____ [2 marks]

c   Describe a test to show that oxygen is produced.

_____

_____ [2 marks]

d   Use the diagram of the experiment to help answer these questions.

   i   What is the dependent variable?

   _____ [1 mark]

   ii What is the independent variable?

   _____ [1 mark]

e The next diagram shows the results.

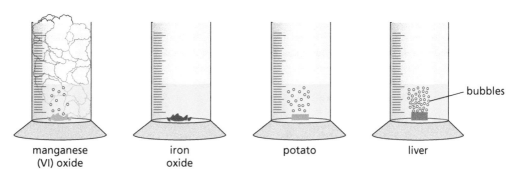

manganese (VI) oxide   iron oxide   potato   liver   bubbles

i Identify the best catalyst. Give a reason for your answer.

_____

_____ [2 marks]

ii Which substance would you **not** recommend as a catalyst for this reaction?

Give a reason for your answer.

_____

_____ [2 marks]

f Anatasia is worried that the results are not valid because the investigation was **not** a fair test. Why does she think this?

_____

_____ [2 marks]

Self-assessment

· ·

Tick the column which best describes what you know and what you are able to do.

What you should know:	I don't understand this yet	I need more practice	I understand this
Metals react in similar ways, but some are more reactive than others			

	I can't do this yet	I need more practice	I can do this by myself
Information about the reactivity of metals with oxygen, water and dilute acids is used to put them in the reactivity series			
A more reactive metal will displace a less reactive metal from its compounds			
Displacement reactions can be used to compare the reactivity of different metals			
The rate of a reaction increases as temperature increases; when the concentration of reactant particles increases; and when the surface area of the reactants increases			
Chemical reactions only occur when the reactant particles collide with each other			
An increase in temperature or concentration increases the rate of reaction because reactant particles collide more often			
Breaking a reacting solid into smaller pieces gives more surface area for the particles of the other reactant to collide with and react			
A catalyst increases the rate of a reaction			

You should be able to:	I can't do this yet	I need more practice	I can do this by myself
Make observations and measurements			
Describe patterns seen in results			
Use a range of materials and apparatus and control risks			
Select ideas and produce plans for testing based on previous knowledge, understanding and research			
Choose the best way to present results			
Describe patterns (correlations) seen in results			
Interpret results using scientific knowledge and understanding			

If you have ticked 'I don't understand this yet' or 'I can't do this yet' or mostly 'I need more practice', have another look at the relevant pages in the Student's Book. Then make sure you have completed all the questions in this Workbook chapter and the review questions in the Student's Book. If you have already completed all the questions ask your teacher for help and suggestions on how to progress.

Teacher's comments

....

Test-style questions

....

1. Aiko investigates the reaction between magnesium and hydrochloric acid.

- She adds a 1 cm length of magnesium to 25 cm^3 of the acid.

- She measures the time it takes for all the magnesium to react.

- She then repeats the experiment at different temperatures.

The table shows her results:

Temperature	Time for magnesium to react (s)
20	181
30	112
40	49
50	43
60	26

a What is missing from the table?

_____ [1 mark]

b Which temperature shows the fastest reaction?

_____ [1 mark]

c **i** Which data point does **not** fit the pattern?

_____ [1 mark]

ii What should Aiko do with this data point?

_____ [1 mark]

2. Mary and Mia are making lemonade. They add some sugar to make it sweeter.

Mary adds three sugar cubes to her lemonade.

Mia adds the same mass of sugar granules to her lemonade.

a Whose sugar will dissolve first? Give a reason for your answer.

_____

_____

_____ [1 mark]

b What could Mia do to speed up the rate at which her sugar dissolves?

_____

_____ [1 mark]

c Use your ideas about particles to explain your answer to part (b).

_____

_____

_____

_____ [3 marks]

3.

Look at the diagram of the reactivity series. It shows some metals in order of reactivity.

Most reactive

K	potassium
Na	sodium
Ca	calcium
Mg	magnesium
Al	aluminium
C	carbon
Zn	zinc
Fe	iron
Sn	tin
Pb	lead
H	hydrogen
Cu	copper
Ag	silver
Au	gold
Pt	platinum

Least reactive

a Name a metal in the diagram that will **not** react with a dilute acid.

_____ [1 mark]

b Name a metal that will react with cold water.

_____ [1 mark]

c Name **two** metals that react slowly with hydrochloric acid.

_____

_____ [1 mark]

d Complete this equation:

calcium + hydrochloric acid → _____ + _____   [2 marks]

e Use the reactivity series to predict which of the displacement reactions below will take place.

Tick **one** box.

☐ calcium + sodium chloride   ☐ magnesium + iron sulfate

☐ lead + zinc chloride   ☐ copper + tin sulfate   [1 mark]

8.1 Temperature changes in reactions

Learning outcomes

- To understand that energy transfers take place during all chemical reactions
- To identify exothermic and endothermic processes from temperature changes
- To evaluate a method and suggest how to improve it

1. Which device uses a chemical reaction to produce a decrease in temperature?

Tick **one** box.

☐ Self-heating can   ☐ Instant cold pack

☐ Hand warmers   ☐ Car engine   [1 mark]

2. Thermometers are used to measure temperature.

What temperature is this one showing? _____ [1 mark]

3. Complete the table.

Worked Example

Reaction	Start temperature (°C)	Final temperature (°C)	Temperature difference (°C)
A + B	19	27	27 − 19 = +8 (1)
C + D	18	11	11 − 18 = −7 (1)

State which reaction is exothermic. Explain how you can tell.

A + B is exothermic (1) because there is an increase in temperature. (1)   [4 marks]

Remember

To find a difference, subtract the starting temperature from the final temperature. Don't forget the minus sign if the temperature has decreased.

4. Safia and Mia mixed some citric acid with sodium hydrogencarbonate solution.

Show Me

The girls recorded the temperature of the mixture at the start of the reaction. They also recorded the lowest temperature of the reaction mixture.

- Starting temperature 20 °C

- Lowest temperature 7 °C

Describe and explain what happened during the reaction.

Remember

When asked to *explain* your answer you must give a reason for what you have written down.

The temperature decreased by _____,

so _____ was transferred from the

surroundings during the reaction as heat.   [2 marks]

5. Youssef was investigating the temperature changes that take place when magnesium ribbon is added to dilute hydrochloric acid. Here are his results.

Length of magnesium (cm)	Temperature at start (°C)	Temperature at end (°C)
1.0	20.1	20.9
2.0	20.2	21.6
3.0	19.9	22.3
4.0	20.0	23.1

State the conclusions Youssef can make from this data.

_____

_____

_____   [3 marks]

6. Investigating energy changes in chemical reactions

Practical Yun and Safia are investigating the energy changes that take place when calcium and dilute hydrochloric acid are mixed together. During the investigation the students record the temperature at the start and the end of each reaction.

97

Figure **X** is a diagram of Yun's apparatus and Figure **Y** is a diagram of Safia's apparatus.

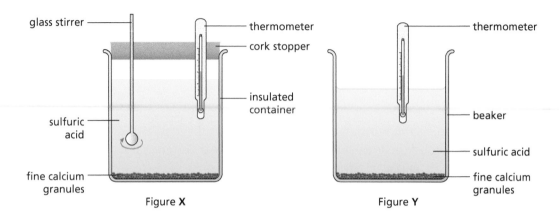

glass stirrer
thermometer
cork stopper
insulated container
sulfuric acid
fine calcium granules
Figure **X**

thermometer
beaker
sulfuric acid
fine calcium granules
Figure **Y**

a Compare the two sets of apparatus. List **two** similarities and **two** differences.

_____

_____

_____

_____ [4 marks]

b Explain why Yun used a cork stopper.

_____ [1 mark]

c Explain which set of apparatus (Yun or Safia's) will give the most accurate results. Give a reason for your answer.

_____

_____ [1 mark]

d Name **two** other pieces of apparatus they will need to use to make sure their results are accurate. Explain your answers.

_____

_____ [2 marks]

8.2 Exothermic and endothermic processes

Learning outcomes

- To determine whether a process is exothermic or endothermic based on the temperature change
- To give some examples of exothermic and endothermic processes

1. Which of the following is an example of an endothermic process?

Tick **one** box.

☐ Freezing   ☐ Condensation

☐ Photosynthesis   ☐ Combustion   [1 mark]

2. Complete the sentences that follow by choosing words from this list.

increase	decrease	endothermic	light		
	exothermic	combustion	heat	energy	

_____ processes transfer energy to the surroundings – often as heat – and

cause a temperature _____ .

_____ processes transfer energy from the surroundings, often as

_____ .   [4 marks]

3. Priya and Lily are investigating the energy changes that take place during dissolving.

Challenge

They follow this method:

- Pour 10 cm³ of water into a boiling tube.

- Record the temperature of the water.

- Add 0.5 g of ammonium chloride and stir.

- Record the temperature when the ammonium chloride has dissolved.

Here are their results:

Start temperature in °C	20.0
Final temperature in °C	18.5

Priya thinks that if they repeat the experiment using 9 cm³ water the temperature change will stay the same, but Lily thinks it will decrease further.

Explain why Lily is correct.

_____

_____ [2 marks]

4. The word equation for photosynthesis is:

Challenge

water + carbon dioxide → glucose + oxygen

The word equation for respiration is:

glucose + oxygen → carbon dioxide + water

Use your scientific knowledge to explain which process is exothermic and which is endothermic.

_____

_____

_____ [4 marks]

8.3 Energy transfers in changes of state

Learning outcomes
- To give examples of endothermic processes
- To explain what happens during melting and evaporation in terms of energy transfer

1. Draw four lines to match each change of state with its energy change.

Change of state   **Energy change**

| Melting |

| Exothermic |

| Freezing |

| Condensing |

| Endothermic |

| Evaporating | [2 marks]

2. Explain how sweating cools the skin.

Show Me

Energy transfers from the skin to sweat as

_____ . The sweat _____

leaving the skin cooler. [2 marks]

> **Remember**
> When answering an 'explain' question you need to give a reason for your answer.

3. Oliver and Mike put some ice into a container and heat it.

They record the temperature every minute. This graph shows their results.

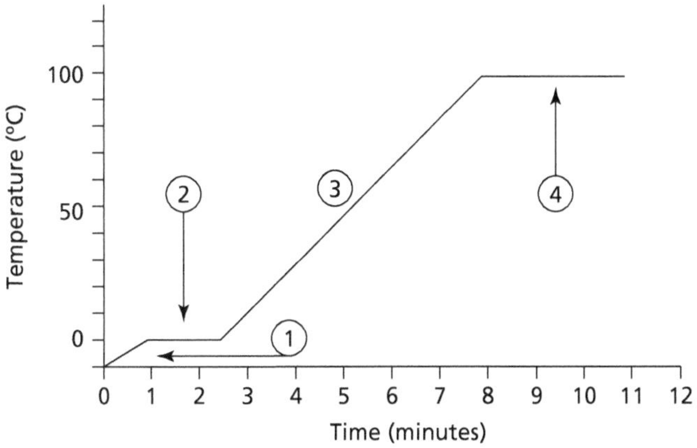

Use the graph to answer these questions.

a What was the temperature of the ice at the start? _____ [1 mark]

b At what temperature did the ice melt? _____ [1 mark]

c Explain what is happening to the particles during step 2.

_____

_____ [2 marks]

d Name the type of energy change taking place during step 4. Give a reason for your answer.

_____

_____ [3 marks]

Self-assessment

Tick the column which best describes what you know and what you are able to do.

What you should know:	I don't understand this yet	I need more practice	I understand this
During chemical reactions, energy can be transferred to or from the surroundings			
The transfer of thermal (heat) energy to or from the surroundings causes the temperature to change			
Exothermic reactions transfer energy to the surroundings as heat, causing the temperature of the surroundings to increase			
Endothermic reactions transfer energy from the surroundings to the reaction, causing the temperature of the surroundings to decrease			
Combustion is an example of an exothermic reaction. Photosynthesis is an example of an endothermic reaction			
Melting and evaporation are endothermic processes			
Evaporation causes cooling because the particles that escape are the fastest-moving ones with the most energy			

You should be able to:	I can't do this yet	I need more practice	I can do this by myself
Decide which apparatus to use and assess any hazards in the laboratory			
Describe patterns seen in results			
Interpret results using scientific knowledge and understanding			
Draw conclusions			
Explain results using scientific knowledge and understanding – communicate this clearly to others			

If you have ticked 'I don't understand this yet' or 'I can't do this yet' or mostly 'I need more practice', have another look at the relevant pages in the Student's Book. Then make sure you have completed all the questions in this Workbook chapter and the review questions in the Student's Book. If you have already completed all the questions ask your teacher for help and suggestions on how to progress.

Teacher's comments

Test-style questions

1. The diagram shows a burning candle.

a Describe what happens to the wax as the candle burns.

_____

_____ [2 marks]

b When wax burns is the reaction exothermic or endothermic? Give a reason for your answer.

_____

_____ [1 mark]

c Safi has drawn a sketch graph to show how the temperature of wax changes as the candle burns.

Explain why there is a plateau in the graph.

_____

_____ [1 mark]

2. Chen is investigating endothermic and exothermic reactions. He mixes different chemicals together and records the temperature changes.

a Complete the table of results.

Chemicals	Start temperature (°C)	Final temperature (°C)	Temperature change (°C)	Endothermic or exothermic
Sodium hydroxide and sulfuric acid	18	21	+3	
Ammonium chloride and water	18	13	−5	
Magnesium and hydrochloric acid	18	25		

[2 marks]

b Predict what will happen to the temperature change if Chen increases the mass of magnesium used but keeps all other conditions the same.

_____ [1 mark]

c Chen is wanting to make a cool pack to keep his lunch fresh.

i Which chemicals could he use?

_____ [1 mark]

ii Describe **two** things he could do to make the pack more effective.

_____

_____ [2 marks]

3.

Practical

Gabriella wants to find out which hydrocarbon fuel releases the most energy when it is burned. She heats equal volumes of water using a known mass of each fuel.

She sets up the apparatus shown in the diagram on the right.

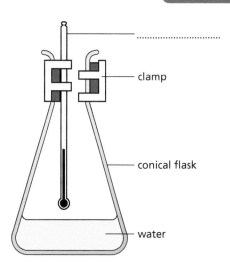

....

clamp

conical flask

water

a Complete the diagram by adding the missing two labels. [2 marks]

b During the investigation, Gabriella records some data in a table.

Write in the missing headings in the table.

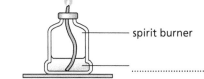

spirit burner

....

Name of fuel	Mass of fuel at start (g)			Temperature of water at end (°C)

[2 marks]

c Look at the diagram. Does all the heat produced go into raising the temperature of the water? Give a reason for your answer.

_____

_____ [1 mark]

d When a hydrocarbon fuel burns it produces carbon dioxide and water. Write a word equation for the reaction.

_____ [2 marks]

e After completing her experiments Gabriella decides to extend her investigation to include wood, which is a solid. Describe how she should modify her apparatus.

_____

_____ [2 marks]

Physics

9.1 Charge and electrostatics

Learning outcomes

- To describe electrostatics using the idea of charge
- To investigate how charges affect each other

1. Complete the sentences that follow by choosing words from this list.

| carbon | negative | positive | neutral | atoms | neutrons | protons | electrons |

Everything is made up of _____ . Inside each atom, there is a nucleus

containing particles with a _____ charge. Outside the nucleus are particles

called _____ which have a _____ charge.   [4 marks]

2. Describe how electric charge can be moved from one object to another.

Worked Example

Rub one object against another object. (1) The friction between the objects can cause some of the electrons on the outside of atoms to move from one object to the other. (1)

Electrons have a negative charge, so one object loses electrons and the object becomes positively charged. (1) The other object gains electrons and becomes negatively charged. (1)   [4 marks]

Remember

Particles called protons have a positive charge and particles called electrons have a negative charge. But atoms are electrically neutral, meaning that each atom has the same number of electrons as it has protons.

The electrons are on the outside of the atoms, so they can move.

3. Look at the diagrams below. Each diagram shows a different pair of charged objects that are brought close together, but not touching.

A

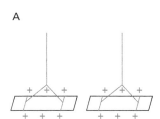

B

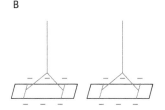

C

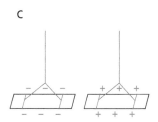

The table is to show the observations of this investigation. Complete the table using the words 'attract' and 'repel'.

Diagram	Observation
A	
B	
C	

[1 mark]

4. We can calculate the net (overall) charge on an object if we know how many positive charges and how many negative charges there are. Look at the diagrams below. Calculate the sign and size of the net charge on each object.

(a)

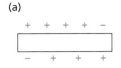

(b)

(c)

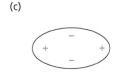

(d)

Object **a** has seven positive charges and two negative charges, so the net charge is:

$7 \times 1 + 2 \times (-1) = 7 - 2 \,(1) = +5 \,(1)$

Object **b** has: _____ [2 marks]

Object **c** has: _____ [2 marks]

Object **d** has: _____ [2 marks]

5. Describe what happens if a proton is brought close to another proton.

Protons are _____ so two protons _____ . [2 marks]

6. Explain why an electron is attracted to a proton.

_____

_____

_____ [3 marks]

Lily uses the apparatus in the image below to investigate how a charged insulator can affect the behaviour of everyday objects. Look at the diagram carefully and answer questions **7** to **9**.

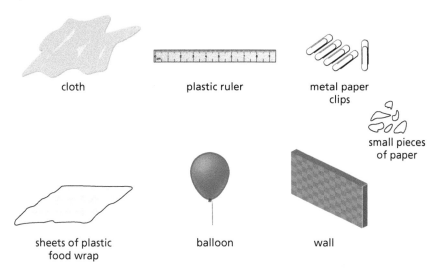

cloth   plastic ruler   metal paper clips

small pieces of paper

sheets of plastic food wrap   balloon   wall

7. Describe and explain a method Lily can use to charge an object.

_____

_____   [2 marks]

8. Look at the table. The first column shows objects that Lily gives charges to. She then brings each charged object slowly closer to each of the objects in the second column.

Complete the table by describing what you would expect Lily to observe for each object.

The first row has been completed for you.

Charged object	Test object	Observation
Ruler	Small pieces of paper	Pieces of paper are attracted to the ruler
Sheet of plastic food wrap	Another sheet of plastic food wrap	
Balloon	Wall	
Ruler	Metal paper clips	

[3 marks]

9. Lily wants to test how tap water is affected by a charged object.

Practical

a Describe how Lily can do this.

Remember to include:

- which items of apparatus Lily needs

- a method.

_____

_____

_____ [3 marks]

b Describe what you would expect Lily to observe.

_____

_____ [2 marks]

c Explain the observation.

_____

_____

_____ [3 marks]

10. Static electricity is used by car manufacturers to help paint some parts of cars. The paint is sprayed onto the car parts using a device called a spray gun. The spray gun and paint are given a positive electric charge.

a What charge should be given to the car parts to attract the particles of paint?

_____ [1 mark]

b Explain how using electric charge in this way means that paint can be applied to parts of cars that are hidden from view.

_____

_____

_____ [3 marks]

11. Carlos investigates electric charge. He uses the apparatus shown in the diagram below.

Challenge

Practical

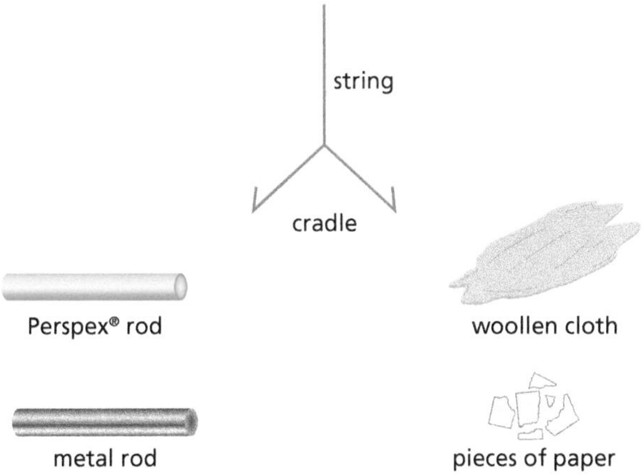

string

cradle

Perspex® rod

woollen cloth

metal rod

pieces of paper

Remember
Light travels in straight lines. Attraction between electrically charged charges can change their direction of motion.

There are two of each type of rod – Perspex® and metal. Carlos rubs one Perspex rod with a woollen cloth before placing it in the cradle.

a Explain how Carlos should use the apparatus to make the Perspex® rod in the cradle move without touching the rod, the cradle or the string.

_____

_____

_____ [3 marks]

b Explain why Carlos cannot move a metal rod using the same method.

_____

_____

_____ [2 marks]

12. Read the article about Aiko and then answer the questions.

> Aiko was out walking with friends when a heavy rainstorm started. They took shelter near some trees. Her friends crouched down near to the ground but Aiko stayed standing up. She felt her hair starting to stand up on end and then there was a bright flash of light and a loud bang. She felt hot and very scared, but was not hurt. The lightning had struck a tall tree nearby.

a Explain why Aiko's hair stood up soon before the lightning strike.

_____

_____

_____ [3 marks]

b Use your knowledge of electric charge to describe what happened when the lightning struck the tree.

_____

_____ [2 marks]

c Remember what you have learned about energy. Write a flow chart showing the energy transfers that took place when the lightning struck the tree.

[4 marks]

9.2 Series and parallel circuits

Learning outcomes
- To measure current using an ammeter
- To make and draw series and parallel circuits
- To investigate the differences between series and parallel circuits, including the effects on current
- To interpret and predict the behaviour of different circuits

1. Name each of the components in the diagrams below.

Choose your answers from this list.

| resistor | switch | cell | battery | diode | ammeter | voltmeter | lamp |

_____

_____

_____

_____

_____

[5 marks]

2. Draw the symbol for an ammeter.

[2 marks]

3. Describe what the current in a circuit represents.

Show Me   The current represents a _____ .   [1 mark]

Look at the diagrams below showing four circuits and then answer questions **4** and **5**.

a

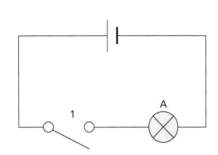

b

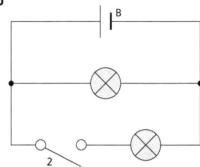

c

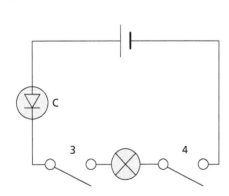

d

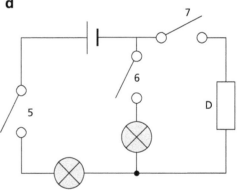

4. Complete the table that describes the four circuits in the diagrams above.

Circuit	Series or parallel?	Switches closed to make lamps light	Components
a	series		**A** =
b		2	**B** =
c			**C** =
d			**D** = resistor

[9 marks]

5. Look at the diagram of **circuit b**. Yuri reads in a textbook that the current in each branch of a circuit adds up. This would mean that:

current from the cell = current in upper branch + current in lower branch

a What component should Yuri add to the circuit to measure the current?

_____ [1 mark]

b Should this component be connected in series or in parallel?

_____ [1 mark]

c Yuri only has one of these components. Describe how Yuri could use this component to test the statement in the textbook.

_____

_____

_____ [3 marks]

6. The electric circuits in houses include circuit breakers or fuses. These are devices that stop the current if there is a fault.

Challenge

a Describe how a circuit breaker or fuse can stop a current from flowing.

_____

_____

_____ [3 marks]

b Explain why a circuit breaker or fuse must always be connected in series with the devices connected to a circuit. What could happen if a circuit breaker was connected in parallel with a device that developed a fault?

_____

_____

_____ [3 marks]

9.3 Explaining current in circuits

Learning outcomes

- To explain how cells (batteries) affect current
- To investigate the effect of changing the length of resistance wire in a circuit
- To understand the effects of a range of different components in circuits
- To use a range of components to make useful circuits

1. Look at the diagrams below. For each diagram, complete the circuit so that all the components are complete and connected. For **circuits a** and **b**, use just *one straight* connecting wire. For **circuits c** and **d** complete the components. Note that you do not need to draw the switches as closed.

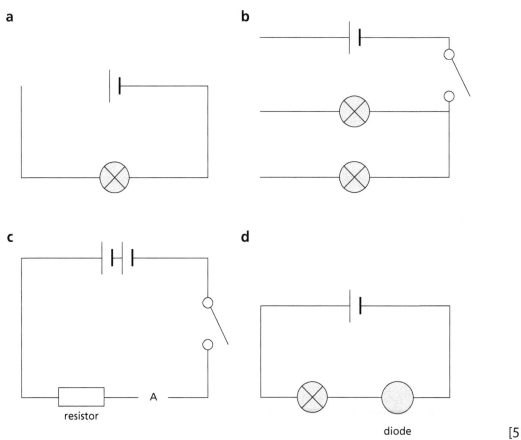

a

b

c

resistor

A

d

diode

[5 marks]

2. Explain why every circuit must be complete for it to work.

Show Me

Current is a flow of _____. (1)

It must be able to flow from one _____. (1)

If a circuit is not complete the current _____. (1) [3 marks]

3. What does a cell or battery provide usefully when it is connected in a circuit?
Tick **one** box.

☐ It makes new charges.

☐ It produces heat energy that is transmitted along the wires and warms the circuit components.

☐ It transfers chemical energy to electrical energy to 'push' the current.

☐ It stores kinetic energy of the charges in the circuit. [1 mark]

4. If a cell is used for a long time eventually it will stop working. Suggest why this is. Choose the *best* answer from the list.

☐ The cell pushes all the charges out of itself, so it has no more charge left to provide.

☐ The cell transfers all the chemical energy it stores to electrical energy, so none is left.

☐ The cell pushes all the electrons out of itself, so it only has positive charges left.

☐ The cell still contains electrical energy, but must now be connected up the other way around to use it. [1 mark]

5. Describe what a diode does in a circuit.

_____

_____ [2 marks]

Questions **6** to **9** are about circuits and components. Look at the diagrams below which show four different circuits. Then answer the questions.

a

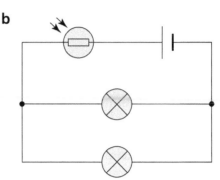

b

c

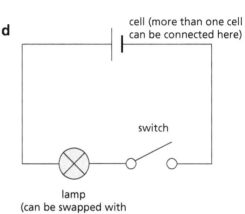

d

cell (more than one cell can be connected here)

switch

lamp
(can be swapped with other components)

6. Look at **circuit a**. Which switch (1 or 2) should be closed to make the lamp light?

_____   [1 mark]

7. Look at **circuit b**. The component is a light-dependent resistor.

a Describe what happens when the amount of light falling on the light-dependent resistor changes.

_____

_____   [2 marks]

b Suggest a use for a circuit like this.

_____

_____   [2 marks]

8. Priya builds **circuit c**. She finds that the lamps do **not** shine brightly.

a Suggest what Priya could add to the circuit to increase the current and make the lamps shine more brightly.

_____ [1 mark]

b Use your knowledge of energy and energy transfers to explain why the lamps will shine more brightly when this component is added.

_____

_____

_____

_____ [3 marks]

9.

Practical

Challenge

Chen builds a circuit to measure the size of the current in different components. He draws a diagram of his circuit. (**circuit d** on page 118). The lamp can be replaced with other components. More than one cell can be connected in the circuit.

Here are Chen's results:

Component	Number of cells used	Current in amperes
Lamp	1	0.6
Resistor	2	1.6
20 cm length of resistance wire	1	1.0
50 cm length of resistance wire	1	0.4

a Chen has made a mistake in his diagram. Which component has he left out of his diagram?

_____ [1 mark]

b Here is a copy of the diagram. Add the missing component to the circuit.

cell

switch

lamp
(can be swapped with
other components)

[2 marks]

c Chen cannot write a conclusion based on his results because he has **not** controlled one of the variables. State which variable is **not** controlled, and describe what Chen needs to do to make sure he can compare all his results properly.

_____

_____

_____

_____ [2 marks]

d Suggest why the current measured for the 50 cm length of resistance wire is lower than the current for the 20 cm length.

_____

_____

_____ [2 marks]

9.4 Voltage

Learning outcomes

- To measure voltage using a voltmeter
- To explain how a voltmeter affects current
- To know that a cell transfers chemical energy to electrical energy
- To plan and do an investigation to find the best materials for a cell

1. Complete these sentences by choosing the correct words. Circle the *best* answer.

a For electrons to move around a circuit, **energy** / **current** must be supplied.

b We can measure the amount of (the answer from part **a**) at a point in a circuit. We call this measurement the **current** / **voltage**.

c We can use **an ammeter** / **a voltmeter** to measure this quantity. [3 marks]

2. A measuring device can be analogue or digital.

a Explain why a digital device is often more accurate than an analogue device.

A digital device shows _____ .

An analogue device requires the user to _____ . This is

not _____ . [3 marks]

b Explain why it is important to check that a measuring device shows a value of zero (0.00) before switching a circuit on and making measurements.

_____

_____ [2 marks]

3. Look at the diagram of a cell (right).

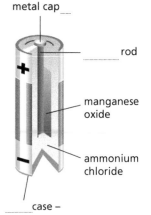

- **a** Complete the diagram by writing in the missing labels. [3 marks]

- **b** Explain why it is dangerous to cut a cell or a battery open.

  _____

  _____

  [2 marks]

4. The figure below shows some energy transfers that take place when a cell is connected to a lamp.

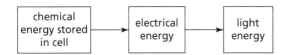

The diagram does **not** show all the energy released by the lamp. State what energy transfer is missing from the diagram.

_____ [2 marks]

The circuits below are used in questions **5** to **10**.

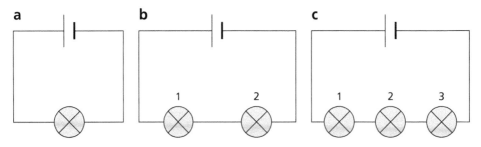

5. The voltage measured across the cell is the same for each circuit. Use ideas about energy to explain why this is.

_____

_____ [2 marks]

6. Compare **circuit b** with **circuit a**. Ahmed measures the voltage across the first lamp in each circuit. He finds that the voltage across the first lamp in **circuit b** is about half the voltage across the lamp in **circuit a**.

Explain why this is.

_____

_____ [2 marks]

7. We can write the relationship in **circuit a** and **circuit b**:

voltage across lamp 1 in **circuit b** < the voltage across the lamp in **circuit a**

Which of the following statements is true about the voltage across lamp 1 in **circuit c** compared to lamp 1 in **circuit b**? Tick the *best* answer.

☐ voltage across lamp 1 in **circuit c** < voltage across lamp 1 in **circuit b**

☐ voltage across lamp 1 in **circuit c** = voltage across lamp 1 in **circuit b**

☐ voltage across lamp 1 in **circuit c** > voltage across lamp 1 in **circuit b** [1 mark]

8. Ahmed takes **circuit b** apart and makes a new **circuit d**. The two lamps are now connected in parallel with each other and the cell.

a Draw a diagram of **circuit d**. Label each of the lamps as '1' or '2'.

[3 marks]

b Which of these statements shows the correct relationship between the voltage of the cell and the voltages across lamps 1 and 2? Tick **one** answer.

☐ V across cell = V lamp 1 = V lamp 2

☐ V across cell = V lamp 1 + V lamp 2

☐ V across cell > V lamp 1 > V lamp 2

☐ V across cell < V lamp 1 < V lamp 2 [1 mark]

9. Blessy uses **circuit d** to test the relationship between the voltage across the cell and the voltages across each lamp.

Write a method for Blessy's investigation. Assume that Blessy has only one voltmeter.

_____

_____

_____ [3 marks]

10. Blessy extends her investigation to include **circuit b**. She also decides to measure the current in each lamp and the cell. Here is a partly completed table of results.

Challenge

Circuit	Current in amperes			Voltage in volts		
	Cell	Lamp 1	Lamp 2	Cell	Lamp 1	Lamp 2
B	3.0	3.0		1.5	0.75	
D	3.0	1.5				

Use your knowledge of current, voltage and also series and parallel circuits to complete the table of results with the values you would expect to observe. [6 marks]

Self-assessment

Tick the column which best describes what you know and what you are able to do.

What you should know:	I don't understand this yet	I need more practice	I understand this
All atoms contain positive charge in the nucleus and negative charge in the surrounding electrons			
Electrons can be lost or gained by friction			
Insulators may gain a net overall charge because the charges cannot move within the insulating material			
Like charges repel. Unlike charges attract			
A current is a flow of electric charge carried by electrons			

	I can't do this yet	I need more practice	I can do this by myself
Current is measured in amps using an ammeter			
The current is the same all round a series circuit			
In a parallel circuit the current splits, but the total current remains the same			
The flow of current can be controlled using different components in a circuit			
Increasing the length of resistance wire in a circuit decreases the current			
A cell contains chemicals that react to generate a voltage			
Voltage is a measure of the energy supplied to electrons by a cell			
A voltage is needed to make electrons flow round a circuit			
If cells are connected in series, their voltages are added together			

You should be able to:	I can't do this yet	I need more practice	I can do this by myself
Plan investigations based on previous knowledge, understanding and research			
Decide which measurements and observations to make			
Decide on which apparatus to use			
Draw and build series and parallel circuits			
Use explanations to make predictions			
Make conclusions using scientific knowledge and understanding			

If you have ticked 'I don't understand this yet' or 'I can't do this yet' or mostly 'I need more practice', have another look at the relevant pages in the Student's Book. Then make sure you have completed all the questions in this Workbook chapter and the review questions in the Student's Book. If you have already completed all the questions ask your teacher for help and suggestions on how to progress.

Test-style questions

1. The diagrams show four different circuits **a** to **d**.

a

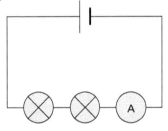

b

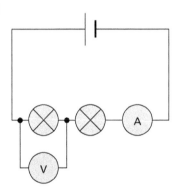

c

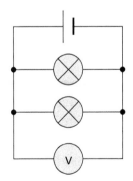

d

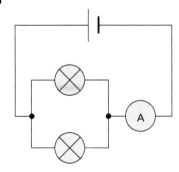

a Which circuit could be used in an investigation to measure *both* current *and* voltage?

_____

[1 mark]

b Name the components used to measure:

i voltage

_____

ii current

_____

[2 marks]

c Which circuits include lamps connected in series?

_____

[2 marks]

d Which circuits include lamps connected in parallel?

_____

[2 marks]

2.

Practical

Angelique investigates the current in a circuit, and how it changes when the length of a resistance wire is changed. Look at the circuit diagram. The symbol for the variable resistor represents the resistance wire.

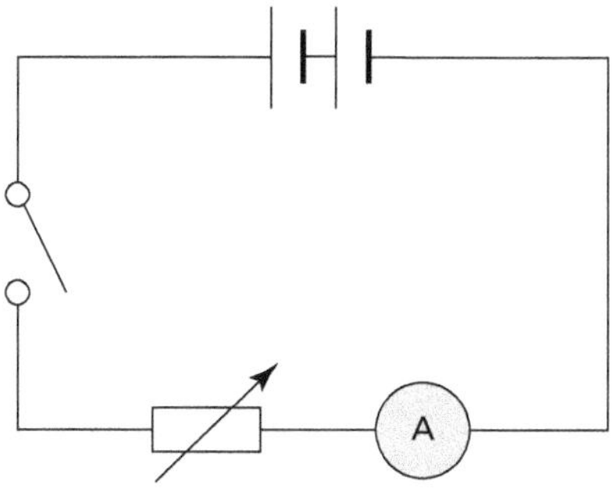

a Predict what you would expect to observe when the length of resistance wire is changed.

_____

_____

[1 mark]

Look at the table of Angelique's results.

Length of resistance wire in cm	Current in A
10	2.0
20	2.2
30	0.8
40	1.2

b Do these results fit with your predicted observations? Describe how the results do (or do not) fit your prediction.

_____

_____ [2 marks]

c Which, if any, lengths of resistance wire should Angelique test again? Explain your answer.

_____

_____ [2 marks]

3. Draw the circuits for the following investigations or uses.

a Two lamps in parallel with a cell and one switch that controls both lamps.

[3 marks]

b Three cells in series with a resistor and a switch.

[3 marks]

c This diagram shows two lamps and a light-dependent resistor. Complete the circuit diagram so that the lamps are in parallel with a battery containing two cells, and the light-dependent resistor switches both lamps on and off.

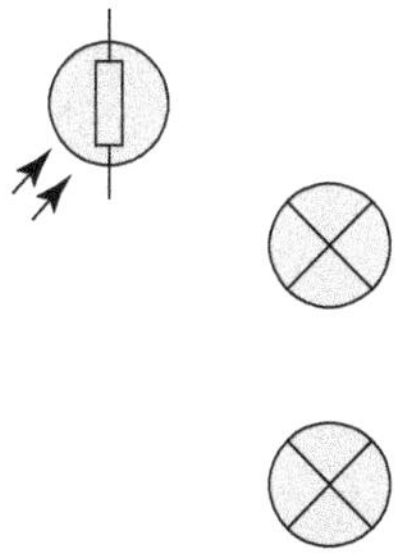

[3 marks]

4. a Explain what an electrical current is.

_____

_____ [2 marks]

b Describe how a diode can be used to change a current.

_____

_____ [2 marks]

5. a Explain how a cloth can be used with an insulator to demonstrate the movement of electrostatic charge.

_____

_____

_____ [3 marks]

b Explain why we need to use an insulator in part a, not a conductor.

_____

_____

_____ [3 marks]

The diagram shows a Van de Graaff generator.

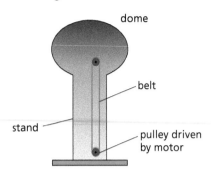

c Which part of the Van de Graaff generator works like a rubbing cloth?

_____ [1 mark]

d What type of charge builds up on the dome when the Van de Graaff generator is switched on?

_____ [1 mark]

e When someone standing on an insulator touches an uncharged dome and waits while the dome is charged up, their hair stands up on end. Describe and explain what is happening with the electric charges.

_____

_____

_____

_____ [3 marks]

10.1 Turning effect of a force

Learning outcomes

- To know how forces can cause objects to turn around a pivot
- To understand the principle of moments
- To use the principle of moments to calculate forces, distances and moments
- To make observations and measurements of turning force experiments

1. Look at the diagrams. Complete each by adding labels from the words in this list.

| force | pivot | distance | load | height |

[4 marks]

2. Complete the equations to show the relationships between **moment**, **force** and **distance from the pivot**.

a moment = —————————— × —————————— [1 mark]

b —————————— = $\dfrac{\text{——————————}}{\text{distance from pivot}}$ [1 mark]

c distance from pivot = $\dfrac{\text{——————————}}{\text{——————————}}$ [1 mark]

3. Complete the table by writing the names and symbols of the units for force, distance and moment.

Force	Distance	Moment
newtons, symbol _____	_____ , symbol _____	_____ _____ , symbol N m

[4 marks]

A turning force of 5 N acts at a distance of 0.5 m from a pivot. Calculate the moment.

moment = force × distance from the pivot

$= 5 \, N \times 0.5 \, m$ (1)

$= 2.5$ (1) N m (1)   [3 marks]

Remember

When you calculate numerical answers you should always:

- write the equation you are going to use
- show your working
- include the units in the answer.

5. Look at these diagrams. Use the values in each diagram to calculate the moments.

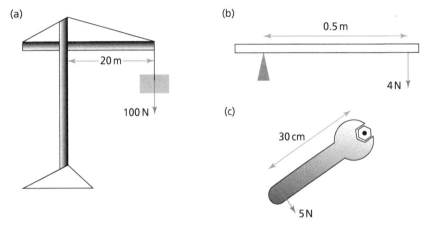

(a)

20 m

100 N

(b)

0.5 m

4 N

(c)

30 cm

5 N

a moment = force × distance from the pivot

$=$ —————— N × —————— m

$=$ —————— N m   [3 marks]

b _____

_____

_____   [3 marks]

c _____

_____

_____   [3 marks]

6. What is the principle of moments?

_____

_____

_____   [2 marks]

7. Look at this diagram.

a Calculate the moment for the left-hand side of the beam.

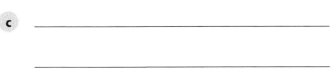

_____

_____

_____ [3 marks]

b Calculate the moment for the right-hand side of the beam.

_____

_____

_____ [3 marks]

c Is the beam balanced? Explain your answer.

_____

_____ [2 marks]

8. Look at these diagrams. All three beams are balanced. Calculate the missing quantity in each beam.

a _____

_____

[3 marks]

b _____

_____

[3 marks]

c _____

_____

[3 marks]

(a)

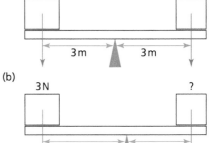

(b)

(c)

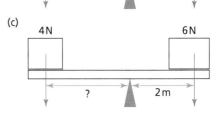

9.

Practical

Look at the diagram. Some devices for weighing vegetables and fruit use a balance like a see-saw. The object to be weighed is placed in the tray on the left-hand side. On the right-hand side, the different weights can be slid to different distances from the pivot.

a Explain why the distance from the pivot to the object being weighed must be measured.

_____

_____ [2 marks]

b What type of variable is the distance from the pivot to the object being weighed? Choose from the list – underline your answer.

| independent | dependent | control |

[1 mark]

c You are given a bunch of bananas to weigh. You are asked to find the weight of one banana but you are not allowed to break the bunch apart. Suggest a method using the see-saw balance to find the weight of one banana.

_____

_____

_____

_____ [4 marks]

d Yuri is given an identical see-saw balance and another bunch of bananas. His result is different from yours.

Suggest **three** possible reasons why Yuri's result is different.

_____

_____

_____ [3 marks]

10. **Challenge** Look at the diagram of a see-saw. Ahmed wants to join in – he weighs 400 N. Explain where Ahmed should sit to balance the see-saw.

Ahmed
400 N

300 N ↓ ←2.0 m→ ←2.0 m→ ↓ 500 N

_____

_____

_____

_____

_____

_____ [6 marks]

10.2 Pressure on an area

Learning outcomes

- To explain that pressure is caused by the action of a force on an area
- To make observations and measurements

1. Complete these equations to show the relationships between **pressure**, **force**, and **area**.

a $$\text{pressure} = \frac{\rule{3cm}{0.4pt}}{\rule{3cm}{0.4pt}}$$ [1 mark]

b $$\text{force} = \rule{3cm}{0.4pt} \times \rule{3cm}{0.4pt}$$ [1 mark]

c $$\frac{\rule{3cm}{0.4pt}}{\rule{1cm}{0.4pt}} = \frac{\rule{3cm}{0.4pt}}{\text{pressure}}$$ [1 mark]

135

2. Oliver investigates the pressure caused by different weights. He has a tray that measures 10 cm × 10 cm on which he places the weights. The first weight he tries is 1 N.

a Calculate the area of the tray in square centimetres and then in square metres, m².

Area = 10 cm × 10 cm = 100 cm² (1)
There are 100 cm in 1 m, so there are
100 × 100 = 10 000 cm² in 1 m² (1)
Area of tray = 100 cm² ÷ 10 000 = 0.01 m² (1)

[3 marks]

> **Remember**
> Prefixes tell us the size of a unit.
> k = kilo-, meaning × 1000, so 1 kg = 1000 g
> m = milli-, meaning ÷ 1000, so $1 \text{ mm} = \frac{1}{1000} \text{ m} = 0.001 \text{ m}$
> c = centi-, meaning ÷ 100, so $1 \text{ cm} = \frac{1}{1000} \text{ m} = 0.01 \text{ m}$

b Calculate the pressure caused by the tray when a 1 N weight is placed on it.

Pressure = force ÷ area

= _____ N / _____ m²

= _____ N/m²

[3 marks]

c Here is a table showing Oliver's results. In the first row, write in the answers to parts **a** and **b**. Then write in the area and calculate the pressure for each of the other rows.

Weight in N	Area in m²	Pressure in N/m²
1		
2		
5		
10		

_____

_____

_____

[3 marks]

3. Blessy has a heavy parcel of weight 300 N that she wants to place on a soft floor. The size of the parcel is shown in the diagram.

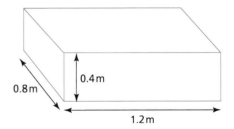

0.4 m
0.8 m
1.2 m

a Which side should Blessy place downwards on the floor to reduce the pressure to a minimum?

_____

[1 mark]

b Calculate the pressure if the parcel is placed on the floor in this way.

_____

_____

_____ [3 marks]

4.

Challenge

A truck weighs 120 000 N. The truck carries a load that weighs 305 000 N. The truck needs to cross an old bridge. The maximum pressure the bridge can withstand is 200 000 N/m². When the truck is fully loaded, one tyre has 0.1 m² in contact with the road surface.

a Work out the pressure if all the weight of truck and its load was placed on an area the size of one tyre's contact patch.

_____

_____

_____ [3 marks]

b Work out the *minimum* number of tyres the truck should have for it to cross the bridge safely.

_____

_____

_____ [3 marks]

c Most trucks have at least two more tyres than they need to satisfy the regulations.

Suggest and explain **one** reason why this is the case.

_____

_____ [2 marks]

10.3 Pressure in gases and liquids

Learning outcomes

- To explain how pressure arises in gases and liquids
- To describe some uses of pressure in gases and liquids
- To explain results using scientific knowledge and understanding – communicating this clearly to others

1. Complete the sentences below by choosing words from this list.

pressure	volume	depth	surface	particles	compressible
increases	decreases	stays the same	incompressible		

Pressure in a liquid is caused by _____ in the liquid pushing against a

_____ . As you go deeper in a liquid, the pressure _____ .

Unlike a liquid, a gas is _____ . This means that if the

_____ of gas decreases, the _____ of the gas increases.

[6 marks]

2. Look at this diagram.

a   Write down the place (A, B, C, or D) where the pressure is highest.

_____   [1 mark]

b   Describe what would happen if a small piece of light wood was placed at point B.

_____

_____   [1 mark]

c   Use your knowledge of pressure to explain your answer to part **b**.

_____

_____   [2 marks]

d   Where is the pressure lowest? Explain your answer.

_____

_____

_____   [3 marks]

3. A piston contains an amount of nitrogen gas. The piston can be moved up or down to different positions, as shown in the diagram.

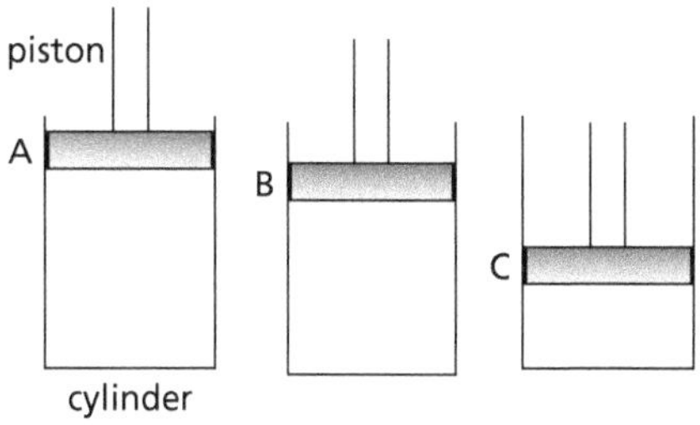

a   In which position is the volume of the gas largest?

_____   [1 mark]

b   Describe as fully as possible what has happened to the volume in position C.

_____

_____   [2 marks]

c   In which position is the pressure of the gas highest?

_____   [1 mark]

4. The pressure in a liquid can be used in hydraulic devices. The diagram shows a hydraulic device that can lift a car.

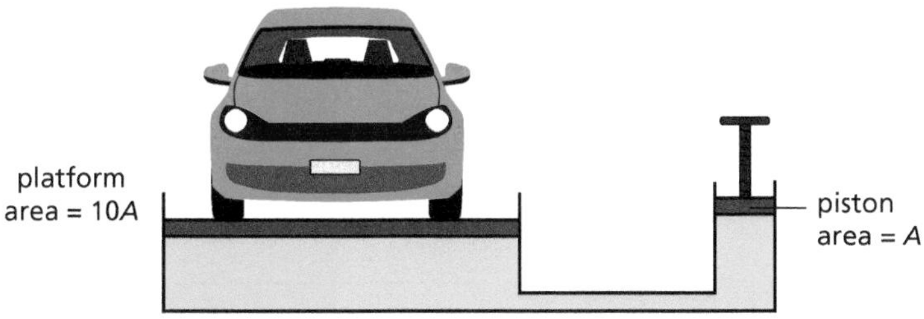

a   Hydraulic devices use liquids, not gases. What property of liquids makes them suitable for use in hydraulic devices, but gases unsuitable?

_____   [1 mark]

b Complete the following sentences by choosing words and phrases from the list. Use each word and phrase once, more than once or not at all.

| pressure | volume | force | equal to | less than | more than |

The pressure in the liquid under the car platform is _____ the pressure in the liquid under the piston.

This means that the _____ pushing up on the platform is

_____ the _____ used to push down the piston.

This makes it easier for a person to lift the car. [4 marks]

c Look at the different areas of the piston and the platform. If a force *F* is used to push down on the piston, what is the size of the force pushing up on the platform?

_____

_____ [2 marks]

5.

Challenge

Anastasia uses a newton meter to measure the weights of different objects. First, she fixes the newton meter to a stand and hangs each object from the meter in air. Then, holding the stand, she carefully lowers the meter and object into a large container of water. She takes a second reading from the newton meter after placing the object in the water.

The table shows Anastasia's results.

Object	Weight in air, in N	Weight after lowering into water, in N
A	5.0	0.0
B	10.0	0.0
C	20.0	7.0

a Describe the results for object A.

_____

_____ [2 marks]

b Explain the results for object B.

_____

_____ [2 marks]

c Explain the results for object C.

_____

_____

_____ [3 marks]

6. Rajiv has a sample of gas in a leakproof container with a piston. He has a measuring device that tells him the volume of the container and the pressure of the gas accurately.

Rajiv changes the volume by moving the piston. He is very careful to make sure the temperature stays constant.

He plots a graph of his results. Use the graph to answer the questions.

Remember

Gases follow a set of relationships between different quantities. Rajiv's graph shows the relationship between pressure and volume – this follows Boyle's Law.

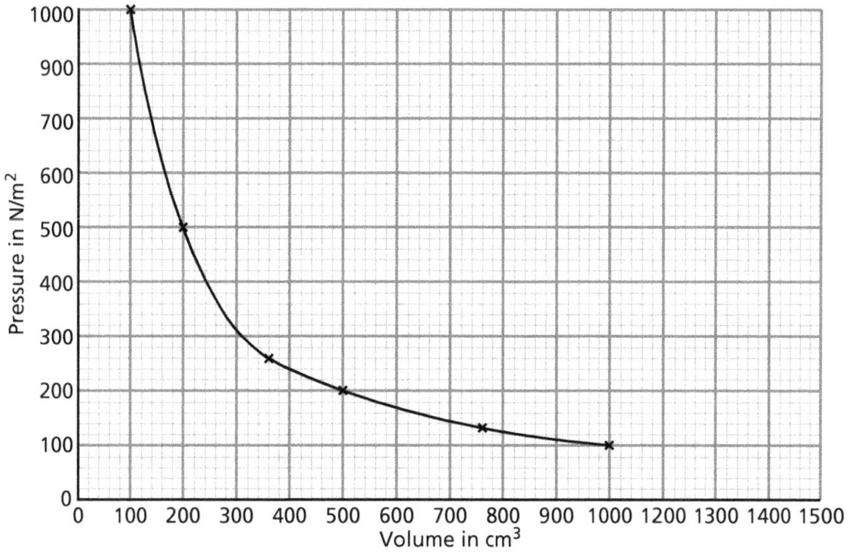

a What is the pressure when the volume is 200 cm³?

_____ [2 marks]

b What is the pressure when the volume is 500 cm³?

_____ [2 marks]

c Explain why it is important to keep the temperature constant.

_____

_____ [2 marks]

141

d Write a conclusion that describes the pattern in Rajiv's results.

_____

_____ [1 mark]

10.4 Density

Learning outcomes
- To define and calculate density
- To investigate and explain the different densities of solids, liquids and gases

1. Write the equation that defines density using the words and units in the list. Write the units in the brackets.

density	volume	mass	kg	m³	kg/m³

_____ (_____) = _____

(_____) divided by _____ (_____)

[6 marks]

2. Thinking about question 1 – what other units we might use for density?

_____ [1 mark]

3. Carlos has a cube of iron that measures 10 cm along each side. He wants to calculate the density of the iron. Carlos puts the cube on a balance and measures its mass to be 7900 g.

Show Me

a Calculate the volume of the iron cube.

_Each side is _____ cm, so the volume is_

_____ cm × _____ cm × _____ cm = _____ [2 marks]

> **Remember**
> Always include the units in the answer to every calculation.

b What is the density of the iron?

_____

_____

_____ [3 marks]

c The density of aluminium is 2.7 g/cm³. Suggest why aluminium, not iron, is used to build aircraft.

_____

_____ [2 marks]

4. The table shows the densities of different materials.

Material	Density in g/cm³	Main properties
Lead	11.3	Strong. Soft. Toxic.
Steel	8.0	Strong. Hard. Non-toxic.
Aluminium	2.7	Strong. Flexible. Non-toxic.
Bamboo	0.3	Strong. Hollow. Can bend a lot.
Fibreglass	2.5	Can be woven to make strong, stiff objects. Can bend a little.
Carbon fibre	1.6	Can be woven to make very strong, stiff objects. Can bend a little.
Nylon	1.1	Can be stretched to make thin fibres. Strong.
Cotton string	1.5	Can be stretched to make thin fibres. Strong.
Polythene	0.9	Can be stretched to make thin fibres. Not very strong.

Look at the diagram. Lead is a metal that was used to make small weights to hold down fishing lines under water.

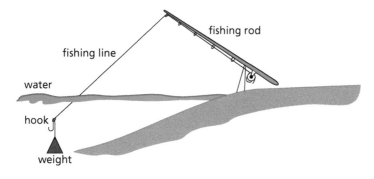

a Suggest why lead was used to make these small weights.

_____

_____ [2 marks]

b Lead is toxic to animals and people. Fishing weights are now made using other materials. Look at the table. Which material in this table would be best to use instead of lead?

_____ [1 mark]

c Explain your answer to part **b**.

_____

_____ [2 marks]

d Which material would you use to make the fishing line? Explain your answer.

_____

_____ [2 marks]

e Fishing rods can be made from many different materials including bamboo, fibreglass and carbon fibre. Suggest why these materials are suitable for making fishing rods.

_____

_____

_____ [3 marks]

5. You can rearrange the density equation to work out other quantities.

Worked Example

a Fresh, pure water has a density of 1.0 g/cm³. A container filled to the top with pure water holds 2.0 litres. What is the mass of water in the container in kg?

1 litre = 1000 cm³, so 2.0 litres = 2000 cm³. (1)

density = mass ÷ volume,

so mass = volume × density = 2000 × 1.0 (1)

= 2000 g = 2.0 kg (1)   [3 marks]

> **Remember**
>
> Check the units you are given and the units you need in the answer carefully.
>
> 1 litre = 1 dm³ = 1000 cm³ = 0.001 m³; and 1 kg = 1000 g
>
> It is often best to convert these units to those you need before doing calculations, but you can also do these conversions at the end.

b Air has a density of 1.2 kg/m³. Calculate the mass of air in a 2.0 litre container.

_____

_____

_____ [3 marks]

c What is the volume of a 1.0 tonne mass of pure water? (1 tonne = 1000 kg)

_____

_____

_____ [3 marks]

6. Look at this table of densities.

Material	Density in kg/m³
Pure water (liquid)	1000
Pure ice (solid water)	917
Sea water (liquid water with added salt and minerals)	1030
Wood (pine)	700

a Is solid (frozen) pure water *more* dense or *less* dense than pure liquid water?

_____ [1 mark]

b Ice floats on liquid water. Use your answer to part **a** to explain why this happens.

_____

_____ [2 marks]

c Predict whether ice will float or sink in sea water. Explain your answer.

_____

_____ [2 marks]

d Explain by referring to density why wood is a good material to use to build a boat.

_____

_____ [2 marks]

7.

Challenge

Practical

In question **6**, you compared the density of different types and forms of water. Refer to the table in question **6**.

a A sample of pure water is heated (but not until it boils). Explain what you would expect to happen to the density of the water.

_____

_____ [2 marks]

b Usually, when a liquid changes state to solid, its density increases. Use your knowledge of solids, liquids and particles to suggest why this occurs.

_____

_____

_____ [2 marks]

c Water is an unusual material because when it freezes, it expands. Describe an experiment you could do to show that water expands when it freezes.

_____

_____

_____ [3 marks]

d When cold sea water is poured into some warm pure water, at first the sea water sinks to the bottom. Explain why this happens and predict what will happen next.

_____

_____ [2 marks]

e Eventually the sea water and pure water mix completely. What can you say about the density of this mixture? Explain your answer.

_____

_____ [2 marks]

Self-assessment

Tick the column which best describes what you know and what you are able to do.

What you should know:	I don't understand this yet	I need more practice	I understand this
You calculate moments by multiplying the force by the distance of that force from a pivot			
Moments are measured in newton metres			
If the clockwise moment and the anticlockwise moment are equal then the objects are balanced			
Pressure arises when a force is applied over an area			
To calculate pressure, use pressure = force ÷ area			
An object with a large area exerts a low pressure			
An object that exerts a high force creates a high pressure			
Pressure is caused by particles hitting surfaces			
As the depth of liquid or gas increases, so does its pressure			
Hydraulics can be used to transfer pressure from one place to another using liquids			
Density is calculated by mass ÷ volume			
Solids are more dense than liquids and gases because their particles are the closest together			
Objects that are less dense than water float on it			
Objects that are more dense than water sink in it			

You should be able to:	I can't do this yet	I need more practice	I can do this by myself
Compare results and evaluate methods used by others			
Use explanations to make predictions and then evaluate these against evidence			
Describe how to obtain reliable results			
Make conclusions using scientific knowledge and understanding			

Describe patterns (correlations)				
Take readings from graphs				

If you have ticked 'I don't understand this yet' or 'I can't do this yet' or mostly 'I need more practice', have another look at the relevant pages in the Student's Book. Then make sure you have completed all the questions in this Workbook chapter and the review questions in the Student's Book. If you have already completed all the questions ask your teacher for help and suggestions on how to progress.

Teacher's comments

Test-style questions

1. Moments, pressure and density all involve measurements of length. Use words from the list to complete the sentences and equations below.

distance	area	volume	force	mass

a **Moments**

The equation is moment = ———————————————— [2 marks]

b **Pressure**

The equation is pressure = ———————————————— [2 marks]

c **Density**

The equation is density = ———————————————— [2 marks]

2. The diagram shows a mass balance that can be used to weigh objects. The object has mass 500 g.

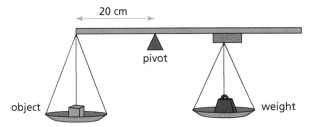

20 cm

pivot

object   weight

a Remember your studies of gravity and weight. State the equation that relates mass and weight.

_____ [2 marks]

b If the acceleration due to gravity, g, is 10 N/kg, calculate the weight of the object.

_____

_____ [2 marks]

c Aiko wants to weigh an object of about 500 g in mass using the balance in the diagram. She only has one weight to use on the balance, with a mass of 250 g. Use the principle of moments to explain how Aiko can use the 250 g weight to balance the object.

_____

_____

_____

_____

_____

_____ [4 marks]

3. Hassan is investigating pressure. He puts a heavy weight on square trays of different sizes. He then puts each tray on an area of very fine sand and sees how deep the tray sinks into the sand.

Practical

a Predict whether or not Hassan should see a pattern in his results. Describe the pattern, if you think there will be one.

_____

_____

_____ [2 marks]

Look at the table of Hassan's results.

Area of tray in cm²	Depth in mm
9	2.0
16	1.6
25	1.3
49	0.5

b Do these results fit your prediction? Explain your answer.

_____

_____ [2 marks]

c Suggest how Hassan could improve his investigation. Explain your answer. (*Hint*: think about how easy it is to get accurate measurements using Hassan's method.)

_____

_____ [2 marks]

d Camels live in the desert and have wide, flat feet. Based on Hassan's investigation and your knowledge of pressure, suggest why camels are well adapted to their environment.

_____

_____

_____ [3 marks]

4. Jamila is a deep-sea diver. She needs to carry a supply of air to breathe underwater.

a Jamila needs to carry as much air as possible, but keep the container small so it does not get in the way. Complete the following sentence using the best word from the list.

atmospheric	low	high

Jamila should carry air at _____ pressure. [1 mark]

b Explain your answer to part **a**.

_____

_____ [2 marks]

c Explain why the air container needs very thick, strong sides.

_____

_____ [2 marks]

d The air container will be heavy because of the thick sides and the amount of air squeezed inside. Explain why the extra weight of the air container will be an advantage for Jamila when she is under water.

_____

_____

_____

_____ [3 marks]

5. To calculate the density of a material, we need to measure the mass and volume of a sample of the material.

Practical

a An object has straight sides and right-angles at the corners. Describe **one** method for determining the volume of this object that does **not** involve liquids.

_____

_____ [2 marks]

b Explain why this method will **not** work for 'irregular' objects (objects with many differently shaped sides or rough surfaces).

_____

_____ [2 marks]

The diagram shows how the volume and mass of an irregular object can be found. The markings on the measuring cylinder are in cm³.

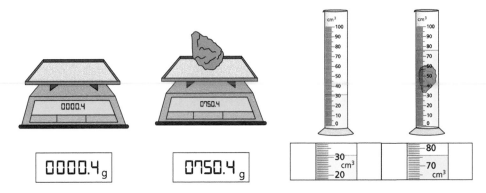

0000.4 g   0750.4 g

c Write down the readings on the measuring cylinder before and after the object is placed in the water.

_____

_____ [2 marks]

d Calculate the volume of the object.

_____

_____

_____ [3 marks]

e What is the mass of the object? _____ [1 mark]

f Use your answers to parts **d** and **e** to calculate the density of the material the object is made of.

_____

_____

_____ [3 marks]

11.1 Energy resources

Learning outcome

• To know the different methods of generating electricity

1. Complete the sentences by choosing words from the list.

| fuels | renewable | non-renewable | biofuels | fossil | nuclear |

Coal, oil and natural gas are all _____ fuels. These fuels cannot be replaced

easily; we say they are _____.

Wind, wave and geothermal energy are examples of _____ energy.

We can also get energy from splitting atoms. This is _____ energy.

[4 marks]

2. A coal power station burns the fuel to generate electrical energy. Describe the energy transfers that take place.

Worked Example

Chemical energy in the fuel is transferred as heat when the fuel burns. **(1)**

The heat is transferred to water, where the thermal energy boils the water into steam. **(1)**

The steam causes a turbine to spin, so the thermal energy is transferred to kinetic energy of the turbine. **(1)**

The kinetic energy of the turbine turns a generator, which transfers the energy to electrical energy in the power grid. **(1)**

Remember

If you are asked to describe what happens in a physical process, make sure that you include:

• where the process happens
• what causes the process to happen
• what changes because of the process.

[4 marks]

3. Look at the diagram of the nuclear power station below.

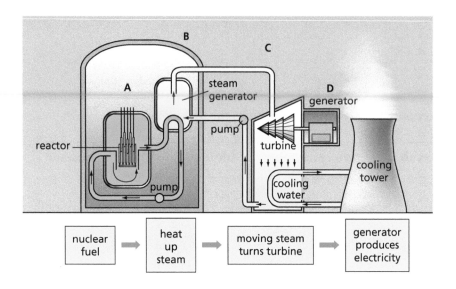

| nuclear fuel | ⇒ | heat up steam | ⇒ | moving steam turns turbine | ⇒ | generator produces electricity |

a Write down the type of energy produced by the fuel at **A**.

_____ [1 mark]

b Describe the energy transfer that takes place between the steam generator **B** and the turbine **C**.

_____

_____ [2 marks]

c Write down the type of energy produced at **D**.

_____ [1 mark]

4. Use a suitable method to list the similarities and differences between the energy transfers in a coal power station and in a nuclear power station. Write short sentences to describe a process or energy transfer for each type of power station.

Show Me

	Coal	Nuclear
Similarities		
Differences		

[5 marks]

5. Describe the functions of the following parts of a power station.

a Turbine

_____

_____ [2 marks]

b Generator

_____

_____ [2 marks]

c Transformer

_____

_____ [2 marks]

6. Look at the image below – it shows a map of part of a country. Use the information on the map to answer the questions that follow.

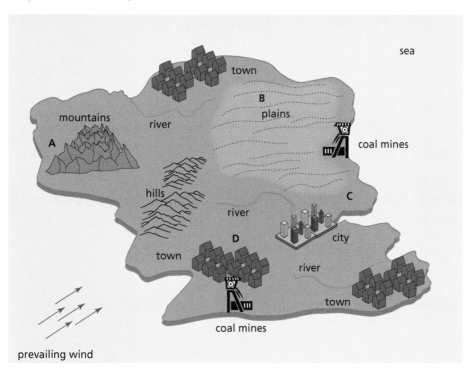

a Explain why coal power stations need to be located near a supply of water.

_____

_____ [2 marks]

b Choose which site, **A**, **B**, **C** or **D**, would be best suited for building a new coal power station. Explain your answer.

_____

_____

_____ [3 marks]

c Suggest why site **D** would **not** be suitable for a nuclear power station.

_____

_____ [2 marks]

d Describe the additional information you would need to choose the best site for a wind farm (a collection of generators powered by the wind).

_____

_____ [2 marks]

7. Hydrogen can be used to produce electricity.

a Name the device in which hydrogen combines with another gas to produce electricity.

_____ [1 mark]

b What is the other gas that is combined with hydrogen to produce electricity?

_____ [1 mark]

c Explain why hydrogen is a renewable fuel.

_____

_____ [2 marks]

d Explain why hydrogen must be stored very carefully.

_____

_____ [1 mark]

8. The UK gets some of its energy from solar power.

a Suggest why the UK meets more of its energy needs from solar power in the summer, compared to in the winter.

_____

_____

_____ [3 marks]

b 'A disadvantage of solar energy is that it is unreliable.' Explain what this statement means and why this is the case.

_____

_____ [2 marks]

9. Switzerland is a country with many mountains but it is landlocked (surrounded by other countries).

Suggest and explain which methods could be used to generate electricity in Switzerland that use water. Why are other methods of electricity production that use water not appropriate?

_____

_____

_____

_____ [4 marks]

10. Companies that extract coal from the ground have suggested a new technology, called 'clean coal', to reduce the harmful emissions from coal power stations. Read the article below from 2014 and answer the question that follows.

Can coal ever be clean?

Coal provides 40 per cent of the world's electricity. It is estimated that burning coal produces 39 per cent of the total emissions of carbon dioxide gas across the world each year. Mined coal not only contains carbon, but also sulfur and nitrogen. 'Clean coal' technology is being developed to reduce the amount of solid carbon and the oxide gases of sulfur and nitrogen which are produced when coal is burned. It does not reduce the amount of carbon dioxide produced.

Burning coal can produce pollution. For example, in December 1952, a thick layer of smog (a combination of smoke and fog) formed over London. It remained there for over three days before it started to clear. The smog contained gases including sulfur dioxide and nitrogen dioxide, as well as large amounts of floating, solid soot (carbon).

During 1953, London had an outbreak of respiratory diseases including emphysema and asthma. Over 12 000 people died from these diseases that year.

Use evidence from the article, as well as your own knowledge, to describe the possible advantages and disadvantages of developing 'clean coal' technology. Include comments on both human health and climate change.

Remember

A respiratory disease is any illness that affects how we breathe.

_____

_____

_____

_____

_____

_____

_____   [6 marks]

11.2 The world's energy needs

Learning outcomes

- To use secondary sources to consider the world's energy needs
- To analyse the secondary sources

1. Complete the sentences by choosing words from this list. You may use each word once, more than once, or not at all.

| **increasing** | **decreasing** | **stayed the same** |

The world's population has always been _____ .

The amount of energy people use has always been _____ .

The amount of non-renewable resources available is _____ . [3 marks]

2. Look at the graph (right) – it shows how a quantity has changed over the past 100 years. It could be one of many different quantities.

a Tick which **three** quantities could be measured using the vertical axis.

☐ Use of renewable energy sources

☐ Total human population

☐ Amount of natural gas remaining to be extracted from the ground

☐ Total carbon dioxide emissions caused by human activity

☐ Total amount of electricity produced from burning oil [3 marks]

b Suggest which piece of evidence from the graph tells you that the quantity **cannot** be the amount of energy produced by nuclear power stations.

The line on the graph does not _____

_____ .

The first nuclear power station _____

_____ . [2 marks]

3. The size of the human population of Earth is affected by many factors. It also *causes* effects on other factors.

a Choose the **three** factors from the following list that *most* affect the size of the human population. Tick the *best* factors.

☐ Amount of energy used by people

☐ Amount of food available

☐ Speed at which mountains form

☐ Availability of suitable places to live

☐ Availability of fresh water

☐ Depth of the sea

[3 marks]

b Choose the **three** factors from the following list that are *most* affected by the size of the human population. Tick the *best* factors.

☐ Amount of carbon dioxide gas emitted

☐ Amount of fuels burned in power stations

☐ Availability of fresh water

☐ Depth of the sea

☐ Amount of waste plastics produced

☐ Speed at which mountains form

[3 marks]

4.

Practical

The table shows the percentage of electricity produced from renewable sources in different countries during 2016.

Country	Amount of electricity produced from renewable sources (%)
Brazil	81.2
Canada	66.4
China	25.8
India	15.0
New Zealand	84.0
Norway	97.9
Saudi Arabia	0.0
United States	15.4

a Suggest why Saudi Arabia produced none of its electricity from renewable energy sources.

_____

_____ [2 marks]

b Suggest **two** reasons why New Zealand produced such a high percentage (84.0%) of its electricity from renewable energy sources.

_____

_____ [2 marks]

c Blessy suggests that a line graph would be the best way to show these data. Say whether you agree with her or not, and explain why.

_____

_____

_____ [3 marks]

d Youssef chooses a bar chart as the best way of showing the data in this question. The figure below shows how Youssef started his chart. Complete Youssef's chart using the rest of the data from the table. Show the values in order from lowest to highest.

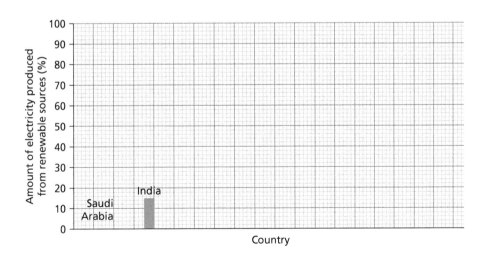

[6 marks]

5. Safia investigates the amounts of different sources of energy used across the whole world during the year 2015. She produces the chart shown below.

Energy use share by resource

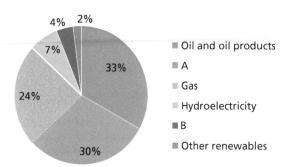

- Oil and oil products
- A
- Gas
- Hydroelectricity
- B
- Other renewables

a Use your knowledge of energy resources to complete the labels for the two segments of the pie chart that are not named.

A _____

B _____ [2 marks]

b How much more energy is used from source A than from gas?

Difference in percentages = % A – % gas

= 30% – 24% (1)

= 6% (1) [2 marks]

c What is the **total** percentage of the world's energy usage that comes from fossil fuels?

Total percentage = _____ + _____ + _____

= _____ % [2 marks]

d What is the **total** percentage of the world's energy usage from renewable sources?

_____ [2 marks]

6. Look at the graph of world energy consumption below. Each shaded colour shows how the amounts of different energy sources used around the world have changed over the past 200 years. The top line of the graph shows the total energy used each year.

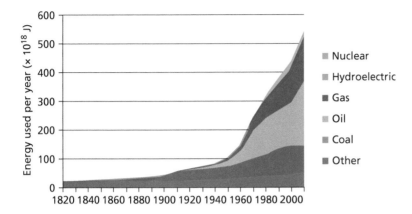

a Name **two** energy sources that must be included in the plotted values labelled 'Other'.

1. _____

2. _____ [2 marks]

b Describe how the **total** energy usage has changed over the past 100 years.

_____

_____ [2 marks]

c State **two** reasons why the total energy usage has changed.

_____

_____ [2 marks]

d Explain why the area marked 'coal' was increasing in size but has started to flatten out.

_____

_____ [2 marks]

7. Water is an essential resource for life. It can also be used in different ways as a source of energy.

Challenge **a** List **five** different methods of electricity production that depend on water.

1. _____

2. _____

3. _____

4. _____

5. _____   [5 marks]

b Explain why untreated water from the sea is **not** suitable for human consumption.

_____

_____

_____ [2 marks]

c Explain why countries in the Middle East have a large number (70%) of the world's desalination plants.

_____

_____

_____ [2 marks]

d It has been suggested that icebergs could be towed from the poles of the Earth for use as supplies of drinking water in the Middle East. Suggest **two** reasons why this could be a better source of drinking water than desalination plants.

_____

_____ [2 marks]

11.3 Conduction, convection and radiation

Learning outcomes

- To describe the different ways that heat is transferred
- To explain how heat transfer can be prevented

1. Draw lines to match the names of the heat transfer processes to the descriptions.

Heat transfer process	Description
conduction	heat causes a liquid or gas to expand so that hotter parts of the liquid or gas rise, and cooler parts sink
convection	heat is transferred but does not require matter to travel through
radiation	heat passes through a substance from particle to particle

[2 marks]

2. Explain the difference between heat and temperature. Which of these quantities can be measured directly? Explain how.

Show Me

Heat is the total amount _____.

Temperature is the _____.

The quantity that can be measured directly is _____.

You can measure this by _____. [4 marks]

3. The three states of matter are solids, liquids and gases.

a List the three states in order so the best conductor is first and the worst conductor is last.

_____ (best)

_____

_____ (worst) [1 mark]

b List the three states in order so the best substance for convection is first and the worst is last.

_____ (best)

_____

_____ (worst)   [1 mark]

4. The figure below shows the inside of an electric kettle.

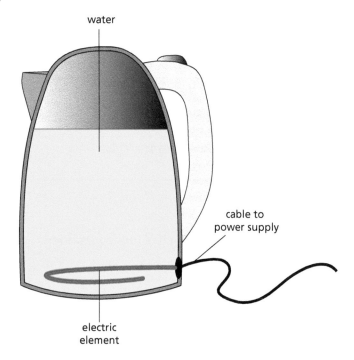

water

cable to power supply

electric element

a What energy transfer takes place in an electric kettle?

_____   [2 marks]

b What type of material should the electric element be made from? Explain your answer.

_____

_____

_____   [3 marks]

c The kettle heats up the water at the bottom. Name the process by which the hotter water moves around the kettle.

_____

[1 mark]

d Draw arrows on the diagram to show the directions in which the water moves as it heats up.

[3 marks]

5. Modern office buildings use large amounts of glass to make the rooms more brightly lit. The glass has a special coating so that the outer surface is shiny but still allows visible light through the glass.

a A room with a large amount of glass can get very hot. What is the process by which heat enters the room?

_____

[1 mark]

b State the source of this thermal energy.

_____

[1 mark]

c Explain why the outer surface of modern office glass has a special coating.

_____

_____

[2 marks]

6. The figure below shows a pan used for cooking.

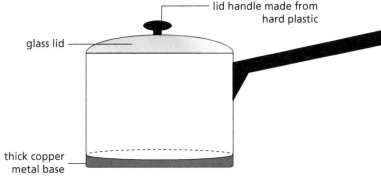

lid handle made from
hard plastic

glass lid

thick copper
metal base

Explain the three labelled features of the pan in terms of how they affect thermal energy transfer.

_____

_____

_____

_____

_____

_____   [6 marks]

7.   Look at the diagrams below which show three experiments involving heat transfer.

Practical

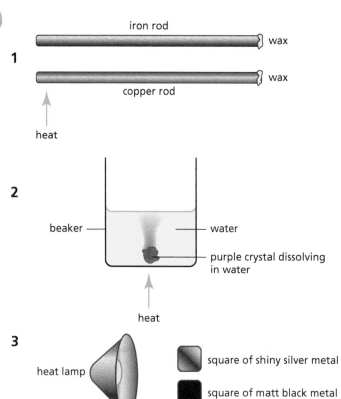

Draw a table for each experiment that shows:

a the name of the heat transfer process that the experiment is designed to show

[3 marks]

b an instruction to make the experiment safer.

[3 marks]

8. In hot weather, your skin sweats. It produces water that rests on the surface of your skin.

Challenge **a** Use your knowledge of heat transfers to explain why sweating helps your body to cool down.

_____

_____

_____ [3 marks]

b If you direct a stream of air over your skin, it can help your body to cool further still. Explain why this happens.

_____

_____

_____ [3 marks]

Self-assessment

Tick the column which best describes what you know and what you are able to do.

What you should know:	I don't understand this yet	I need more practice	I understand this
Electricity is generated using many different methods			
Most methods of generating electricity involve spinning a turbine attached to a generator			
Different factors affect species populations			
Different factors affect the human population			
The increasing human population is creating a higher demand for food, water and energy			

Heat is transferred by conduction, convection, radiation and evaporation			
Conduction occurs in solids and happens when heat is passed from particle to particle			
Convection occurs in fluids and happens when the particles move around because of having more energy			
Radiation is caused by heat being transferred using waves and can occur in a vacuum			
Silver, shiny objects reflect heat well; black, matt objects absorb and emit heat well			

You should be able to:	I can't do this yet	I need more practice	I can do this by myself
Decide whether to use evidence from first-hand experience or secondary sources			
Decide which apparatus to use and assess any hazards			
Choose the best way to present results			
Describe patterns seen in results			
Interpret results using scientific knowledge and understanding			
Look critically at sources of secondary data			
Draw conclusions			
Compare the results and methods used by others			
Explain results using scientific knowledge and understanding			

If you have ticked 'I don't understand this yet' or 'I can't do this yet' or mostly 'I need more practice', have another look at the relevant pages in the Student's Book. Then make sure you have completed all the questions in this Workbook chapter and the review questions in the Student's Book. If you have already completed all the questions ask your teacher for help and suggestions on how to progress.

Test-style questions

· ·

1. The diagram below shows a cross-section through a vacuum flask.

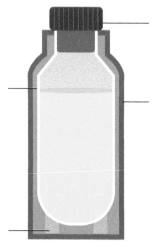

a Use this phrase list to complete the labels in the diagram.

gap (vacuum) between surfaces

thick plastic stopper

shiny metal wall

plastic supports [3 marks]

b Use the phrase list to add the *main* reason each labelled part in the diagram is used. Each answer may be used once, more than once or not at all.

to reduce radiation

to reduce conduction

to reduce convection [4 marks]

c Vacuum flasks are used to keep the thermal energy in hot drinks high for as long as possible. Predict whether or not a vacuum flask can also be used to keep the thermal energy in cold drinks low. Explain your answer.

_____

_____

_____ [3 marks]

2. Nearly all our sources of energy exist only because of nuclear fusion reactions that take place inside the Sun.

a Explain how fossil fuels depend on the energy from the Sun.

_____

_____ [2 marks]

b Suggest why hydroelectricity can only be produced because of energy from the Sun.

_____

_____ [2 marks]

c One resource that we use to produce electricity has **not** come from the Sun.

i Name this energy resource.

_____ [1 mark]

ii Describe **two** advantages of using this energy resource.

_____

_____ [2 marks]

iii Describe **two** disadvantages of using this energy resource.

_____

_____ [2 marks]

3. The diagram below shows Juno, a NASA space probe orbiting the planet Jupiter.

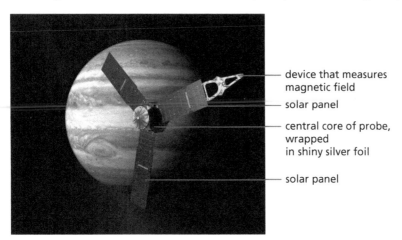

device that measures magnetic field

solar panel

central core of probe, wrapped in shiny silver foil

solar panel

Although it is extremely cold in outer space, the electronics in Juno are very delicate and must be protected from external sources of heat.

a What form of heat transfer can take place in outer space?

_____ [1 mark]

b Suggest where this heat might come from to affect Juno.

_____ [1 mark]

c Describe the feature of Juno that is designed to protect the electronics from heat. Explain how this feature works.

_____

_____ [2 marks]

Juno is designed to use very little energy, but still needs sources of energy to function.

d Describe the feature of Juno that provides a source of energy. What colour is it and why is it this colour?

_____

_____

_____

_____ [4 marks]

e Identify the most important energy transfer that takes place in this feature.

_____

_____ [2 marks]

4. Around the world, coal is the most widely-used fuel in the production of electricity.

a Name the element in coal that is burned to release energy.

_____ [1 mark]

b Describe the energy transfer that takes place when coal is burned.

_____

_____ [2 marks]

c Name the main gas that is produced when coal is burned. Explain why this gas is a pollutant.

_____

_____ [2 marks]

5. Coal is extracted from the ground using one of two different methods:

- Surface mining – the coal is found near the surface and large holes are dug to extract the coal.

- Underground mining – the coal is found deep under the surface and tunnels are dug to extract the coal.

a Suggest and explain which method is cheaper.

_____

_____ [2 marks]

b Suggest and explain which method is more damaging to the environment.

_____

_____ [2 marks]

c The graph below shows the worldwide production of coal over the last 100 years.

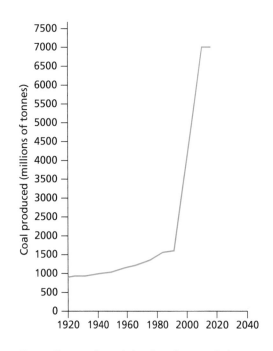

Describe and explain the shape of the graph.

_____

_____

_____

_____ [4 marks]